Legacies of War

Legacies of War

Kathrin Bachleitner

OXFORD
UNIVERSITY PRESS

OXFORD
UNIVERSITY PRESS

Great Clarendon Street, Oxford, OX2 6DP,
United Kingdom

Oxford University Press is a department of the University of Oxford. It furthers the University's objective of excellence in research, scholarship, and education by publishing worldwide. Oxford is a registered trade mark of Oxford University Press in the UK and in certain other countries.

Published in the United States of America by Oxford University Press
198 Madison Avenue, New York, NY 10016, United States of America

British Library Cataloguing in Publication Data
Data available

Library of Congress Control Number: 2025947885

ISBN 9780198982906

DOI. 10.1093/9780198982913.001.0001

Printed and bound by
CPI Group (UK) Ltd., Croydon, CR0 4YY

The manufacturer's authorized representative in the EU for product safety is Oxford University Press España S.A. of Parque Empresarial San Fernando de Henares, Avenida de Castilla, 2 – 28830 Madrid (www.oup.es/en or product.safety@oup.com). OUP España S.A. also acts as importer into Spain of products made by the manufacturer.

FSC
www.fsc.org
MIX
Paper | Supporting responsible forestry
FSC® C013604

To all those who experience war.

Acknowledgements

This book has been a long journey through loss and emotional reckoning. It began with the aim of foregrounding the voices of those who endure war, of giving ordinary people a say. Too often, academic disciplines such as International Relations feel distant from the psychological realities of those who live through political violence. I hoped to help bridge that divide.

At the same time, I remain aware that writing about such experiences within an academic context can never fully capture the depth and complexity of individual suffering. Theories, statistics, and abstract frameworks are often far removed from the lived experience of war. And yet, within those frameworks, there is space for individual voices to resonate. This book is one such attempt. I hope it may encourage others to continue in this direction.

My own scholarly path while writing this book took me through several academic homes. The project began at the University of Oxford, during my time as a Junior Research Fellow at Lady Margaret Hall and the Refugee Studies Centre, supported by the generous funding of the IKEA Foundation. These institutions sharpened my focus on the human dimensions of war. I am grateful to my colleagues and mentors at Oxford, particularly Professor Alexander Betts of Brasenose College, Professor Robin Harding of Lady Margaret Hall, and Professor Stathis Kalyvas of All Souls College. Your insights challenged and inspired me to think beyond disciplinary boundaries and to pursue new forms of understanding.

Two principals of Oxford colleges that I considered home for many years—Lady Margaret Hall and Hertford College—also encouraged me to pursue an individual and original path through academia and beyond: Alan Rusbridger, then Principal of Lady Margaret Hall and now Editor-in-Chief of *Prospect*, and Tom Fletcher, then Principal of Hertford College and now Head of UN OCHA. I learned a great deal from you—about politics, civic leadership, and diplomacy—and your work continues to inspire me.

Finally, I wish to thank Dominic Byatt, my editor at Oxford University Press. Your enthusiastic support for this second book gave me the encouragement I needed to complete it.

In later stages of the project, the University of Salzburg and Ca' Foscari University of Venice offered rich intellectual environments in which to continue writing, teaching, and reflection. The research conducted at Ca' Foscari formed part of the project "The Effects of Conflict and Displacement on National Identity. The Case of Syria," funded by the European Union — NextGenerationEU, National Recovery and Resilience Plan (NRRP) (CUP N.H73C22001310001). I am especially grateful to the Salzburg Centre of European Union Studies (SCEUS) and the Edith Stein Haus for providing a quiet, contemplative office space where much of the manuscript was completed. To the people at these institutions—colleagues, students, and friends—thank you for your support, encouragement, and thoughtful engagement.

This book was also shaped by experiences beyond conventional academic settings, through collaborations with research assistants from conflict-affected areas, and through journeys to, and encounters with people living in Ukraine, Syria, and the Palestinian Territories. These meetings left a lasting impression, and I carry them with me still. Among the many who contributed to this book, I wish to thank in particular Taras Prodaniuk, whose work in data collection, analysis, and visual representation was invaluable.

Beyond my academic and fieldwork experiences, the legacy of war also runs through my own family. My relatives have lived with the impact of the Second World War throughout their lives. My great-grandfather typed out his memories of war in an effort to make sense of the injustices he had witnessed during both world wars. My grandparents spent a lifetime coming to terms with the interruption of their youth, studies, work, and future due to military service, injury, and displacement. My mother grew up in an environment steeped in loss and post-war reckoning, shaped by the memories, silences, and struggles of both her parents and grandparents. Their stories have deeply influenced my own outlook on life.

While this book cannot speak for everyone, it seeks to return attention to those most directly affected by war—those who endure it, and who shape its legacy within their societies. My own biography has made me especially attuned to loss and trauma. I know first-hand the difficulty of finding meaning in the aftermath of profound personal tragedy. The sudden, violent loss of my twin sister, Julia, the most important person in my life, continues to shape me, even fifteen years later. Choosing to live fully, not despite suffering but at times because of it, is, I believe, a superhuman effort. Yet it can open a path to new beginnings, deeper insight, and a more profound embrace of life.

To all those living through the impossible, through experiences that no one should have to endure, but which remain tragically common, especially in times of war, this book is for you. While it cannot fully capture the enormity of war, grief, loss, and trauma, it is a modest attempt to honour your voices and your resilience.

Contents

List of Figures

List of Abbreviations

CIS	Commonwealth of Independent States
EU	European Union
HTS	Hay'at Tahrir al-Sham
IDP(s)	internally displaced person(s)
IR	international relations
IS	Islamic State
NATO	North Atlantic Treaty Organization
OCHA	United Nations Office for the Coordination of Humanitarian Affairs
PLO	Palestine Liberation Organization
UN	United Nations
UNHCR	United Nations High Commissioner for Refugees
UNRWA	United Nations Relief and Works Agency for Palestine Refugees in the Near East
USSR	Union of Soviet Socialist Republics
SDF	Syrian Democratic Forces

Introduction

Framing the Legacies of War

> All of us—great and small, belligerent and neutrals—we must not close our ears to the dire warning of this hour, the threat of such unthinkable horrors. Peace is at hand! As a thought, a desire, a suggestion, as a power working in silence, it is everywhere, in every heart. If each one of us opens his heart to it, if each one of us firmly resolves to serve the cause of peace, to communicate his thoughts and intimations of peace—if every man of good will decides to devote himself exclusively for a little while to clearing away the obstacles, the barriers to peace, then we shall have peace. (Hermann Hesse 1946, *If the War Goes On..*, 36.[1])

This famous plea was written in the wake of the First World War. However, it rings true also in the many instances of war that we witness today: in Ukraine, in Syria, in the Palestinian Territories. War means the threat and the reality of 'unthinkable horrors'. Together, these can give rise to the 'desire' and, consequently, real prospects for peace. The key question, however, is how the traumatic legacy of war can be transformed into the foundation for sustainable peace?

This book is an attempt to answer this question. It explores how peace takes root in the mindset of the people affected by war, even if only as a 'thought', 'desire', or 'suggestion'. From this perspective, this study starts with the assumption that those who experience war retain the agency to transform the conflict. By communicating their 'thoughts and intimations of peace', people hold the power to 'clear away the obstacles and barriers to peace'. In the conceptual terms of this book, individuals can shape *their* 'legacy of war' and its impact on conflict transformation through *their* interpretations and reinterpretations of *their* war experiences.

Under the term 'legacies of war', the book understands the lingering impact of the war experience on people's political attitudes and societal visions.

[1] Reproduced with permission of Canongate Books.

Legacies of War. Kathrin Bachleitner, Oxford University Press. © Kathrin Bachleitner (2026).
DOI: 10.1093/9780198982913.003.0001

It argues that such legacies of war can create opportunities and obstacles for a peaceful transition. The agents of this process are the people who experience war: in this study, Ukrainians, Syrians, and Palestinians. By collecting individual voices from these three peoples currently living through war, these case studies attest that 'peace is indeed everywhere, in every heart'. It is 'a desire and a suggestion' that shines through in people's narratives about their war experiences, their emotional attachments to their countries, and especially in their visions of the future. Yet, as this book highlights, the way these narratives are told, national attachments described, and visions of the future presented, can create barriers or opportunities for transitions to peace.

From an International Relations (IR) perspective, scholars who research wars mainly at the macro level may regard peace in the ideas of individuals—as it is considered in this book—as merely a 'power working in silence'. The theories presented in this book, however, suggest that this silent power can culminate in a force for societal change. But this depends on whether people are able to 'clear away the obstacles and barriers to peace' left behind by the legacy of war. What, then, are these obstacles and barriers? Under what conditions can they be removed? And finally, even if people do manage to remove them, what kind of peace will they get?

What Are War Legacies?

People who experience war come to see their country and its future in a new light. War fundamentally reshapes how individuals understand their society and its political trajectory. Living through armed conflict, people develop new frames of reference about belonging, authority, justice, and peace that endure beyond immediate experiences of violence. This book conceptualizes these transformations as 'legacies of war'. In this study, legacies of war thus refer to the cognitive landscapes of people who live through war.

Existing scholarship typically employs the term 'legacies of war' to denote the material and institutional consequences of conflict—its visible aftermath, including high human death tolls, the destruction of infrastructure, the proliferation of arms and landmines, and economic stagnation.[2] These dimensions are undeniably central to post-war trajectories. Yet, as several studies have noted, war also leaves behind less visible but equally consequential legacies in the form of social meanings, political attitudes, and emotional imprints

[2] For instance, there is an educational advocacy organization called 'Legacies of War' that is dedicated to removing unexploded bombs and landmines in Laos, Cambodia and Vietnam. See: https://www.legaciesofwar.org (viewed 16 September 2024).

that persist over time and, in some cases, across generations (Balcells 2012; Lupu and Peisakhin 2017; Rosoux 2020; Villamil 2021).[3]

This book advances a distinct conceptualization. Rather than treating war legacies as long-term or intergenerational outcomes, it theorizes them as immediate, immaterial transformations in political orientations among those directly experiencing war. War legacies are thus defined here as the evolving political perspectives, national attachments, and future-oriented expectations that emerge during conflict itself.

To examine this intangible legacy of war, the analysis focuses on Ukrainian, Syrian, and Palestinian populations who are currently experiencing conflict. What political ideas do people living through war hold about their country and society, and how do they envision the future?

The Argument

This book examines war legacies by documenting people's subjective experiences of war through their narratives. It argues that the legacies war leaves on societies are expressed in individuals' accounts of conflict, their national attachments, and their visions of the future. The book maps these legacies across three contemporary conflicts—Ukraine, Syria, and the Palestinian Territories—using original data collected through online surveys administered to samples from each population. Finally, it links individuals' interpretations of war, their country, and its future to varying orientations toward peace.

Research Design

This book explores the legacy that war leaves on societies through the narrated experiences, opinions, and attitudes of affected people. With its focus on individuals and their war experiences, the book thus follows a less researched, yet emerging approach within IR that concentrates on the discipline's micro-foundations (Dyrstad 2012; Dyrstad et al. 2011). Such work highlights the importance of local knowledge and lived experiences in sustaining what scholars describe as 'everyday peace' (Firchow 2018; Mac Ginty 2021; Sylvester 2013b). As such, the book adds to a proliferating scholarship within IR that foregrounds deep knowledge of affected actors and localities in processes of negotiating ceasefires and implementing peace agreements (e.g. Acosta 2021; Bar-Tal, Oren, and Nets-Zehngut 2014;

[3] For example, there is a Public Broadcasting Service/TV series entitled 'Legacies of War' that is dedicated to the legacy of the Vietnam War through the experiences of US veterans. https://www.imdb.com/title/tt13209908/ (viewed 16 September 2024).

Chaitin and Avlagon 2025; Dyrstad et al. 2022; Pisano 2022; Sosnowski 2023). In placing the individual and societal dimensions of war front and centre, this book also speaks to IR's notion of human security and its importance for sustainable peace (Newman and Richmond (eds.) 2001). Unlike traditional security concepts which focus on the state's securitization and territorial integrity, human security concepts concentrate on the sociopolitical conditions under which people live during and after war.

In IR, concentrating on the people who live through war, rather than on state actors and their interests, requires looking across to other disciplines for insights and theory-building, including to Social Psychology, Sociology, and Anthropology. These disciplines help to define and research 'legacies of war' as emerging from people's war narratives, their collective identities and nationalistic or patriotic attachments to their countries at war, as well as their political and societal visions of the future. However, specifically for IR, documenting these legacies in different case studies—in this book, in current and ongoing war scenarios in Ukraine, Syria, and the Palestinian Territories—is valuable not only in its own right, but also because such expressions may reveal potential links to peace. Political perspectives, national and social visions, and ideas about the future—shaped by diverse interpretations of war by those who live and suffer through it—can present both obstacles and opportunities for conflict transformation.

This book thus connects the micro level with the macro level by relating people's perspectives to the possibilities and challenges of transitioning from war to peace. With this, the book aims to establish replicable ways in which the findings can travel from the three cases to other scenarios where a legacy of war lingers, with the potential to alter societal and political ideas in the direction of peace.

However, researching war legacies from the bottom up and their potential impact on a macro outcome such as 'peace' immediately warrants a caveat. 'Peace' is a contested concept, and a very vague outcome to observe (Cocker 2014, 85–89). In IR, it is defined predominantly as 'the absence of war', and as such has been researched as part of the dynamic of the war system (Aron 1966, 151; Von Clausewitz 1976). This book's definition follows that trend mainly because its focus is on the legacies of war rather than on peace itself. The book defines legacies of war as the immediate impact that the experience of war has on transformations to peace. For this purpose, it implies that the termination of violence means the beginning of peace. However, of course, such a narrow definition falls short of including the complex processes and conditions that preserve peace rather than transform war (Martín 2005, 45). As all of us lucky enough to enjoy life or periods of life in peace know, peace is much more than the absence of violence. If anything, it is what Johann Galtung famously termed, a 'positive peace' comprising cooperation, integration, and social justice (Galtung 1969, 190). Galtung's positive peace is considered the endpoint of the transition from violence to non-violence; it is what makes peace sustainable in the long run. Legacies of war as they are described in this book will have lasting reverberations on the sustainability of a such defined positive peace. However, for the purpose of this study and because of the nature of its cases (each concerns an ongoing conflict), this book can explore only the first step of the journey towards sustainable peace by highlighting trends that may lead to the absence of violence and thus the appearance, at least, of a 'negative peace'.

The study starts with a theory-building effort that situates the focus on the micro level, on the people who live through wars and conflicts. However, it does also suggest links with the macro level. What, if anything, is the relationship between an individual's narrative of war, their level of patriotic attachment and nationalistic sentiment, and their vision of the future and peace? Can we infer anything from these individual subjective stories from very different war scenarios for the general processes of conflict transformation and peacebuilding? Is there a bridge between individual experiences of war and broader IR questions of war and peace?

Through linking its own definition of war legacies with the interdisciplinary concept of social trauma, the book attempts an answer to these questions by pointing out novel routes between people's perspectives amid war, and the obstacles and opportunities for peace that emerge from these. The book does this by newly defining the legacy that war leaves on society as manifesting in people's interpretations of their war trauma, their attitudes towards their countries at war, and their visions of the future amid conflict. It empirically illustrates such defined war legacies through three case studies consisting of Ukrainian, Syrian, and Palestinian voices. Overall, the book points out replicable ways for IR scholars to document war legacies as people's subjective experiences and perspectives, and their potential consequences for the possibilities of transitioning from war to, at least, the absence of violence.

Structure of the Book

This book maps the legacy that war leaves on societies. It defines this legacy as war's continuing influence on societal outcomes. It suggests that—during wars—such a legacy manifests as the lived experience of war and the subjective interpretation thereof by affected individuals (Chapters 1 and 2).

As such, this book looks at war's legacy from the perspective of personal experience rather than academic distance. It is about how people live through and reflect upon political violence. The book is structured along three dimensions, with each empirical chapter highlighting the voices of affected people from Ukraine, Syria, and Palestine.

Chapter 3 focuses on people's narratives of war. It lets individuals formulate their story and perspective of the conflict they are experiencing. It explores their designations of perpetrators and victims, and their moral assignments of responsibility and suffering. Recounting the war is a way of making sense of the war events and ultimately cements new moral horizons for people and their in-groups.

Chapter 4 explores people's national attachments in times of war, and their feelings towards their country that has slipped into conflict. It asks about their levels of patriotic pride and the reasons why they feel proud/not proud to be part of their national in-group. Chapter 4 hence uncovers how people's national attachments have been affected by war.

Chapter 5 investigates people's ideas of peace and the future of their country. It sheds light on their understandings of their state, its borders, political system, and religious and ethnic make-up. Exploring what kind of future individuals imagine for their country, suggests how an acceptable peace agreement and as a result, a sustainable post-conflict reconstruction process, will look for the people and their society.

While these three chapters highlight different experiences and interpretations of war, the country and its future, Chapter 6 examines these legacies and considers possibilities for conflict transformation. As the theoretical considerations emphasize, people's accounts are not only interesting per se but also contain hints towards the potential for peace. By revisiting people's narratives of conflict, national attachments, and visions of the future, the concluding chapter gauges the opportunities and obstacles these pose for moving away from political violence.

1
Theoretical Framework
War and Its Legacies in International Relations

This chapter lays out the theoretical framework of the book by establishing connections between interdisciplinary concepts, research findings, and the discipline of International Relations (IR). It begins by introducing people as agents of both war and peace. It then defines the legacy of war as the 'social trauma' left on societies. Finally, it presents individuals' subjective interpretations of their traumatic war experiences as a potential explanation for varying social outcomes—particularly the outcome of peace.

Researching War and Peace on the Micro Level: People as Agents of their Legacy of War

War and peace are central to the study of IR. The discipline itself was founded in the wake of the First World War for scholars to understand and thus prevent the outbreak of total war. However, early IR scholars entered the topic on the level of abstractions: they began to talk about states, systems, power balances, and decision makers. For a long time, explaining war and peace became the exclusive realm of realist theories, with scholarship on war distinguished by Kenneth Waltz's three images (Waltz 1959). According to the first image, the locus of war is found in the nature and the behaviour of man, that is, wars result from the selfishness, misdirected aggressive impulses, or simply the stupidity of foreign policy leaders (Waltz 1959, 16). Instead, the second image concentrates on the state level. State-level analyses consider political, institutional and economic structures, the role of public opinion, and political cultures and ideology when it comes to explaining war and peace (Levy and Thompson 2010, 14–15). The third image concerns the international system. On this level, war and peace are about strategic interests, material resources, military capabilities, and security goals (Freedman 2012). Especially since Kenneth Waltz's neorealist turn (1979), scholars look at the anarchic structure of the international system, the number of major powers, the distribution

Legacies of War. Kathrin Bachleitner, Oxford University Press. © Kathrin Bachleitner (2026).
DOI: 10.1093/9780198982913.003.0002

of economic and military power among them, and patterns of alliances. In all cases, the focus is on the macro rather than the micro level.

However, at least since the turn of the century, research has shifted away from global systemic trends towards decision-making processes within states or the nature of states to determine why wars break out and continue (Mearsheimer 1995). 'War is a social institution, and it is waged by societies, not only by states' (Gray 2012, 9). More and more IR analyses of war thus consider the role of social groups when examining war and a potential transition to peace (Levy and Thompson 2010, 14–15). Especially, when with the end of the Cold War, the traditional scenarios of interstate wars gave way to civil wars, the discipline became more focused on actors and their motivations (Levy and Thompson 2010; Wimmer 2014, 177–178). Nevertheless, war and peace in IR are still studied predominantly on the level of elites, and their interests, in other words, on the macro level. What is missing from many IR accounts is the micro level, that is, the focus on those who experience and live through war—the people. Their real-life experience of war is often lost in IR's rather technical and abstract analyses.

While many IR theories and their explorations of war and peace leave people out of the spotlight, a handful of critical IR studies began to explore war from the perspective of the people who experience it. Among them, first and foremost feminist IR scholars stress putting the 'ordinary of war' (Sylvester 2012, 490, see also Sylvester 2013a and b) into focus and exploring different experiences of war among individuals in general, and women in particular (e.g. Elshtain 1987; Scarry 1985). War, they argue, must be thought of 'as experience, as something ordinary people observe and suffer physically and emotionally depending on their locations' (Sylvester 2012, 483). To study war as experience requires that human bodies come into focus as units that have war agency while being prime targets of war violence and war enthusiasm. Ordinary people, in the perspective of feminist IR scholars, thus have agency as spectators or participants in wars. This experience in their view destroys established social meanings: 'little in social and political life goes untouched by war' (Brighton 2011, 101–105). In this definition, the key elements of war are 'its actual mission of injuring human bodies and destroying normal patterns of social relations' (Sylvester 2012, 484). In the feminist IR scholarship, individual people and their experiences of war thus become a primary focus.

War looked at from the bottom up, as injuring people and destroying their social realities, has also entered the research interest of several IR constructivist scholars. Karin Fierke (2004), for instance, in 'Whereof We Can Speak', is focussing on the social experience of war. She addresses how war causes people trauma that leaves them without the means of expressing verbally what they and others in the community have been through emotionally. Building on Wittgenstein's characterization of trauma as an experience 'whereof we cannot speak', she points out cases where victims are unable to speak of something, and as a result, the community cannot be re-established, and its history recorded. With this, nevertheless, Fierke emphasizes meanings and practices that highlight the power of individuals' experiences with war (Fierke 2012).

Along similar lines, scholarship on 'everyday peace' foregrounds how peace is continously produced, negotiated, and sustained through the ordinary practices, relationships, and choices of people living with conflict, thereby challeging elite-centric understandings of peacebuilding (Firchow 2018; Mac Ginty 2021). The concept of human security represents another step in bringing such bottom-up perspectives to the fore within IR. It places individuals, their rights, and development at the centre of security. Unlike traditional security concepts that focus on the securitization of the state and its territorial integrity, human security concepts focus on the sociopolitical conditions in which people live. As such, they lay bare the internal dynamics of conflict that can threaten to prolong political violence (Newman and Richmond 2001, 6). In contrast to traditional approaches, the emphasis is not anymore on decision makers and their rationality in the face of war, but rather on society and perceptions of identity, values, and emotions triggered by the experience of violence (Cockell 2001, 12–18).

Next to critical IR scholars, sociologists and anthropologists long studied wars on the micro level. Within Sociology, there is an emerging scholarship called the 'sociology of war'. This body of research emphasizes the importance of nationalist ideologies and cultural framings when explaining war and the transition to peace (Wimmer 2014, 175). Sociological works thus also help to turn the focus away from strategic decision-making towards the people and their experiences of war. They specifically emphasize cultural frames that define collective sentiments and perceptions central to wars: honour, shame, victory, defeat, friend, and foe (Wimmer and Min 2009). Sociologists also make central the notion of nationalism, as both a consequence and a cause of war (Hiers and Wimmer 2013). With this, while reverting the focus towards the people, sociologists, nevertheless, predominantly look at the collective and social frameworks when researching people's experiences of war.

Anthropologists, on the other hand, attempt to open up the boxes of these collectives. Embedded within social contexts, they conduct an 'ethnography of war' among the people. Traditionally, this endeavour concerned the observation of soldiers; but recently, scholars have begun to include ordinary individuals and their 'everyday' during wars (Käihkö 2022, 677–679). With this, ethnographers add nuance to collectives. They focus on individuals and their diverse voices and stories, thus gathering different opinions within highly polarized war contexts dominated by social frames that paint a very black-and-white picture of in- and out-groups, friends and foes. By focusing on the grass roots, ethnography can shed light on how people think and feel; researchers can not only give a human face to war but also differentiate collective narratives that all too often work like brooms that simplify as they sweep (Käihkö 2022, 677–683). In the wake of polarized collective framings of war, ethnographers now differentiate the picture by shedding light on individuals and their diverse war experiences.

Following these interdisciplinary works, this book places the focus on the micro level. It regards people as central to war, not just as decision-makers and militaries, but as individuals and groups: they can be drivers, perpetrators, victims, or bystanders in wars. It takes collective feelings and emotions triggered by war trauma seriously, shedding light on the psychological undercurrents that operate in the lives of communities when they experience war. People, in this book, are considered agents of war and its legacy. Their actions, experiences, and interpretations are crucial for war and its potential transformation to peace.

Legacies of War as the Social Trauma Left on Societies

The psychological and emotional mechanisms that operate within individuals and societies affected by wars are best described by the interdisciplinary concept of trauma. Trauma is first and foremost in the research focus of psychologists, yet sociologists and IR scholars also picked up on it. While psychologists examine the impact of trauma on individuals, sociologists and political scientists emphasize the social dimension of trauma and its impact on societal outcomes.

Within psychology, trauma is a term used when an individual experiences a tragic event that lies outside any ordinary frames of reference. In the psychological definition, trauma has the following ingredients (McCann and Pearlman 1990): (1) it is sudden, unexpected, or non-normative; (2) it exceeds the individual's perceived ability to meet its demands; and (3) disrupts the individual's frame of reference and other central psychological needs and related schemas. From this definition, it follows that trauma completely overwhelms the individual. However, individuals do not only passively experience victimization through trauma, but they also actively fight back. Strategies to fight trauma include 'denial', 'reversal', 'projection', and 'depersonalization'. While 'denial' silences and avoids the trauma, 'reversal' means to revert the trauma into a valuable experience. 'Projection' involves blaming or scapegoating others, while 'depersonalization' displaces the threat onto another source or rationalizes it. However devastating trauma is to the individual, and whichever the response to cope with it, the experience of trauma always also signifies a new beginning. The world as the person knew it is shattered, and hence must be rebuilt and assigned new meanings (Smelser 2004, 44–46).

Furthermore, not only individuals but also groups can experience trauma. Trauma, in its sociological rather than individual, psychological form, was first defined by Kai Erikson: 'By collective trauma I mean a blow to the basic tissue of social life that damages the bonds attaching people and impairs the prevailing sense of commonality'. For a collective to suffer trauma, it means that its cohesive identity is ruptured, 'the "we" no longer exists as a connected pair or as a linked cell in a larger communal body'. In other words, social identity or group belonging changes, in what can be described as 'a gradual realization that the community no longer exists as an effective source of support and that an important part of the self has disappeared' (Erikson 1976, quoted in Alexander 2012, 9). Similar to individual trauma, social trauma means a total rupture of established social meanings and frames. In a more formal definition of 'cultural trauma', Smelser (2004, 44) describes it as 'an event or situation which is (a) laden with negative affect, (b) represented as indelible, and (c) regarded as threatening a society's existence or violating one or more of its fundamental cultural presuppositions'. Wars, without doubt, are such traumatic events that abruptly rupture the social order and shatter existing forms of societal meaning and belonging.

The experience of trauma in individuals and societies therefore destroys pre-existing meanings, yet also poses a stark opportunity for change and transformation. Several authors point to how the traumatic experience of estrangement can become the basis for a new commonality. In what Erikson (1978) calls a 'gathering of the wounded', society members are united by a sense of being set apart from those who have not suffered. They, however, now share a distinct and novel set of perspectives because of their experience. As a result, they gather a changed sense of self, a changed way of relating to others, and a changed worldview (Schick 2011, 1840). The earth-shattering experience of trauma thus becomes an opportunity to start a new, revise social meanings, rekindle societal foundations, and generate a new sense of group belonging.

However, while this reconstruction for the traumatized individual unfolds as a coping and survival mechanism, in the case of collectives, rekindling the community and its social meaning is a deliberate effort (Alexander 2012; Bell (ed.) 2010; Edkins 2002 and 2003; Hutchison 2016). Irrespective of the gravity of the event, social trauma is not objectively harmful to a collective like individual trauma is to a person. Instead, and even in the face of the most tragic happenings like conflict and wars, the 'injury to the collectivity' must be socially constructed (Alexander 2012, 14). Collective trauma always needs to be framed as having abruptly and harmfully affected a group. This process happens if a larger group of people shares and promotes the belief that they were subjected to trauma (Alexander 2012, 14, 30). To that end, they articulate claims about meaning-making from the trauma experienced in the public sphere. This brings the process of interpretation into the spotlight: People gain agency vis-à-vis the formation of social trauma. However, their interpretations are influenced by public framings of traumatic events, and, as such, can be politically manipulated.

While collective framings of trauma guide how people and societies interpret traumatic events such as wars, these interpretations in turn have an impact on social outcomes. 'Through processes of telling and retelling, trauma can translate into a phenomenon that helps to constitute identities and cohere forms of political community' (Hutchison 2016, 70). With their articulations of trauma, people thus do not only define the rupture that the traumatic event posed for the group but also a pathway for how the social group could be mended in the future (Olick 1998 and 2005; Saito 2006). 'For, indeed, it is the meaning trauma attains—rather than the "actual" experience itself—that helps to constitute forms of collective identity and political community' writes Emma Hutchison (2016, 71). In that sense, the concept of social trauma describes precisely what in this book is called a legacy of war: the experience of war, its subjective interpretation, and the lingering effects on social outcomes that follow. The next section reviews the predicted social outcomes in the wake of war trauma.

War Trauma Ruptures Social Meaning and Reconstitutes Communities

While scholars across the disciplines all agree that the experience of war trauma ruptures pre-existing social meanings, thereby triggering a profound transformation in the self-identifications of individuals and groups, they disagree on the direction and consequences of this change. If trauma, by definition, signifies a rupture of established personal and social ties—as well as individual and collective identities—the central question becomes how these ties and identifications are reconstituted in the aftermath of violence. It is through this process of reconstitution that the legacy of war is forged, with far-reaching implications for individuals' personal and political behaviour and, ultimately, for trajectories of conflict or peace.

Scholars have advanced two main explanations for the impact of trauma on individuals and collectives, particularly their social behaviour, and possible links with peace. Concerning individuals, social psychologists show that the experience of trauma can have both negative and positive effects on people, making them more vengeful or more cooperative (Antoniou et al. 2020; Noor et al. 2017; Zitek et al. 2010). Sociologists and political scientists, particularly scholars working within IR and Conflict Studies, mapped these ideas onto communities (Bell (ed.) 2010; Dyrstad 2012; Edkins 2003; Fierke 2004; 2012; Lerner 2020). They propose that collective trauma can lead to polarization or increased solidarity and national cohesion between groups (Bar-Tal et al. 2014; Hutchison 2016; Hunt 2010; Vollhard and Bilewicz 2013). As a result, it can trigger renewed vicious cycles of violence or novel social bonds that lay the foundations for peace. Let us review each of their propositions in turn.

First, social psychologists explain the impact of trauma on individuals in two ways: on the one hand, post-traumatic growth theory suggests that individuals, after experiencing extremely challenging circumstances, undergo positive transformations, such as developing more meaningful relationships with others and a desire for cooperation. Studies confirm that these individuals show a greater appreciation of life and its value, as well as more social empathy. In other words, they experience emotional growth (Bauer et al. 2016; Hartman and Morse 2018; Tedeschi and Calhoun 1996).

On the other hand, experiencing hardship can lead to post-traumatic stress in which the person retreats from social relations, feels isolated, silences the past, or repetitively and anxiously relives it. This reaction has been defined as a central component of post-traumatic stress disorder (PTSD) frequently observed in trauma victims (Nemeroff et al. 2018). Victimhood resulting from traumatic events, in this view, becomes a central and often harmful part of identity: victims are associated with feelings of rage, resentment, and grief. As a result, social psychologists observe that those who experienced trauma become vengeful, and often the next perpetrators in situations of political violence (Bar-Tal et al. 2013; Vollhard and Bilewicz 2013; Zeitzoff 2018). Trauma victims, in this strand of the scholarship, thus gain a 'bad reputation' for their readiness for conflict and violence.

Scholars of Conflict Resolution in particular picked up on these negative effects of traumatic experiences. With the help of Social Psychology's theory of interpersonal relations (TIR), they show how violence increases in-group favourability and out-group competition (Bar-Tal et al. 2013; Vollhard and Bilewicz 2013). In situations of actual and perceived victimization post-trauma, differences between groups become more salient—escalating and protracting conflict (Bar-Tal et al. 2014; Mitzen 2006). This is because traumatic experiences strengthen group ties, social bonds, and collective identity among those who have become victimized. Michel Foucault called this phenomenon 'the solidarity of the shaken' (Foucault, quoted in Edkins 2003, 5), and, even earlier, Ernest Renan pointed to the utility of victimhood for building a nation: 'Suffering unites' (See Ernest Renan's famous speech, 'What Is a Nation?', 1882). However, while 'shared suffering' creates trust and altruism among in-group members, at the same time it deepens the divides with out-groups, thus exacerbating cleavages, anger, and resentment between them. In a vicious cycle of revenge, more violence follows.

However, traumatic experiences do not necessarily lead to conflict participation and more violence (Canetti et al. 2015; Zeitzoff 2018). In the interdisciplinary literature, victimhood has also become associated with other outcomes. It can trigger any new form of political attitudes and behaviour (Bachleitner 2024; Balcells 2012; Bellows and Miguel 2009; Chaitin and Avlagon 2025) and as such can create social trust and cooperation (Bauer et al. 2016) as well as increased support for peace (Acosta 2021, 682; Nussio et al. 2015). Taking these findings together, the contributions about the impact of different victimhood experiences on political behaviour remain ambivalent as to whether exposure to wartime violence begets further violence or promotes more peaceful attitudes among those affected by wars.

The same ambivalence runs through the IR scholarship. Most IR scholars attest to polarization during conflicts (Hardin 1995; Wood 2008). As a result of political violence, cleavages become more salient (Kalyvas 2006; Laitin 2007), be they based on social inequality (Collier and Hoeffler 1998; Cederman and Gleditsch 2013) or differences in the narration of myths (Kaufman 2001). Whatever their source, these differences all escalate and protract conflict (Bar-Tal 2013; Mitzen 2006). Under such polarized dynamics, collective identities tend to break apart into smaller, competitive units (Rotberg 2004; Ayoob 2007; Bates 2008). Prominent historical examples are the violent fragmentation of Yugoslavia into several ethnically defined states (Peterson 2002; Kalyvas and Sambanis 2005), or the strict separation between Israelis and Palestinians as a result of protracted conflict between them (Bose 2007; Rotberg (ed.) 2006). Most works on the conflicts in Syria and Iraq tell a similar story: once the authoritarian grip was loosened, fighting broke out along diverse substate lines (Hokayem 2017; Phillips 2013 and 2015). The IR scholarship thus also observes that violence begets more violence through polarization, competition, political manipulation, and fragmentation of different groups.

Most IR works have traditionally reiterated the idea that traumatic events have a predominantly negative impact on societal behaviour and collective identities. Political violence in their studies not only changes but 'hardens' identities and social meanings, producing a vicious contest of 'othering' triggered by victimization through traumatic events. However, such findings are largely based on studies that centre on the elite level and on the political manipulation of collective identities in pursuit of particular war-related interests.

By contrast, a smaller body of IR scholarship has begun to examine identity change during and after war at the micro level. Prominently, the work of Karin Dyrstad (Dyrstad (2012); Dyrstad et al. 2022 and 2011) demonstrates that ethno-nationalism does not necessarily intensify in the context of ethnic civil war. Similarly, Rumelili and Todd (2018, 3–8) show that, when attention shifts to discursive justifications articulated by individuals who have experienced war, a 'range of sensible claims to belonging' emerges that extends well beyond the polarizing dynamics typically identified at the elite level (Todd 2018, 88). As a result, identity boundaries may narrow or expand, and the meanings associated with particular groups may be reinterpreted in more exclusive or inclusive terms (see also Todd 2018a). These processes open multiple pathways, leading not only towards polarization and renewed conflict, but also towards conflict transformation and peace.

Taken together, scholars across disciplines agree that the direction of societal change in the aftermath of collective trauma remains open. While trauma invariably disrupts established social meanings and identities, 'in some cases national trauma results in enhancing a sense of unity within a society, there are other cases in which collective traumas have fragmenting effects' (Neal 1998, 31, quoted in Smelser 2004, 44). Trauma may therefore undermine or consolidate collective identities, often doing both simultaneously.

What scholars do converge on is that experiences of trauma and victimhood exert a profound influence on both individual and collective identities, as trauma entails a radical search for meaning and the reconfiguration of social ties. Yet whether these processes lead to 'post-traumatic growth' or 'post-traumatic stress', to 'social empathy' or 'revenge', and to community-building or destruction remains uncertain. The crucial issue, therefore, is not whether trauma transforms societies, but under what social and political conditions the legacy of war contributes to sustainable peace rather than renewed conflict.

Different Interpretations of War Trauma, Diverse Consequences for Peace? Narratives of War, National Attachments, and Visions of the Future

When the social outcomes of war trauma range from conflictual to peaceful behaviours, attention turns to how this experience is interpreted and approached. Social psychologists refer to this as the process of 'working through' trauma. Working through involves confronting traumatic events through interpretation and reinterpretation. As sociologists have shown, these events are not recounted as fixed facts, but rather as socially and emotionally constructed experiences. Multiple interpretations can thus emerge, be established, or even be actively socially manipulated. This interpretive process grants individuals and societies a measure of agency in shaping the legacy of war, highlighting how diverse interpretations can lead to different societal outcomes.

The subjective interpretation of events—including, but not limited to, wars—and their effects on social outcomes is the focus of a large and interdisciplinary body of scholarship on 'narratives'. In this literature, narratives are seen as central to the creation of identity, values, purposive action, and goals (Bachleitner 2021; Chaitin and Avlagon 2025; Monroe 2011; see also the Ontological Security scholarship in IR, e.g. Bachleitner 2023; Steele 2008). Narratives are, by definition, stories about the past (Wertsch 2008); they are subjective interpretations composed of plots and characters (Cruz 2000). The narrators may be individuals, social groups, or the state and its political representatives. Importantly, interpretations vary not only by narrator but also by social context and the time at which an event is recalled. Narrating a given experience is always a reconstruction, shaped by fluid and contemporary frames.[1]

Narratives, especially those of war, are therefore never static or neutral accounts of past events. Rather, they represent a 'meta-code' that arises between emotional experience and the effort to confer meaning (Berenskoetter 2014, 269; Hutchison 2016, 43). They form the 'basis on which [...] the nature of shared reality can be transmitted' (White 1987, 1), offering a rich source of insight. As Monroe (2011, 323) writes, narratives 'illuminate the cognitive landscape and help us understand how people see themselves, how they see others'. Chapter 3 of this book accordingly treats individuals'

[1] See a burgeoning interdisciplinary literature on 'collective memory' spanning Psychology, Sociology, Anthropology, the Political Sciences (e.g. Bachleitner 2021; Confino 1997; Olick et al. (eds.) 2011; Olick and Robbins 1998).

war narratives as both a valuable source of empirical data and an analytical lens, illuminating their cognitive landscapes, and how they perceive their in- and out-groups.

In this respect, narratives are both constructed and constructive. They constitute actor, action, and purpose, holding constitutive, productive, and normative power for the present and future. They create meaningful courses of action appropriate to the acting subject while also serving as sources of moral justification (Berenskoetter 2023).

First, regarding their constitutive function: narratives form identities, for individuals and for groups. They do so by evoking a past experience in order to generate purpose (Somers 1994). Individuals come into being through the stories told by and about them; likewise, the state and nation become meaningful entities through narratives (Ringmar 1996). Put simply, individuals and groups are the stories they tell about themselves.

Notably, when narratives shape identity, the interpretation of war trauma has the potential to reconfigure national attachment. While it is often assumed that war strengthens national unity, attention to individual and varied social interpretations suggests that warfare does not automatically generate in-group solidarity or national homogeneity. Instead, national attachments may be reshaped by differing social interpretations, and a range of claims to belonging may emerge from the experience of war (Rumelili and Todd 2018, 3–8; Todd 2018, 88). As Chapter 4 of this book illustrates, this process opens multiple pathways through which war may affect national attachment.

Second, regarding the productive and normative function of narratives: the interpretation of war trauma through active narration, on both individual and collective levels, not only fosters new forms of national attachment but also lays the groundwork for future action. As constructors of identity, narratives underwrite action by bestowing social meaning on past experiences and deriving lessons from them (Berenskoetter 2014). Specifically, recounting trauma often yields a moral lesson (Bachleitner and Betts 2024; Vinitzky-Seroussi 2002). In this sense, narratives are deeply normative: they contain 'convictions about the possible'—what a group can or cannot do, and normatively, what it must or must not do (Cruz 2000, 277). Narratives of traumatic events such as war thus carry teleological messages envisioning a different or better future ('progress narratives'; Zerubavel 2003, 14–15). As Chapter 5 shows, diverse interpretations of the same experience may produce distinct normative horizons and thus generate divergent outlooks on the future.

In Social Psychology, this interpretive process is termed 'working through' trauma, in contrast to 'acting out'. Working through entails confronting trauma and reinterpreting it in new ways. Acting out, by contrast, involves repressing trauma, allowing it to resurface in unconscious behavioural patterns. Whereas working through affords agency, acting out renders individuals passive in the face of trauma.

However, working through trauma—particularly when it involves war and morally unambiguous atrocities—is profoundly difficult for both individuals and communities. People often struggle to critically examine their traumatic experiences. Although they seek meaning in the aftermath of disaster, the narratives they construct often fall back on simplistic causal explanations. War trauma narratives, in particular, tend to suppress critical self-reflection. In the process of working through trauma, individuals may revert to moralistic, binary tales of good versus evil, leading to demonization of the other and provoking revenge-seeking behaviour (Schick 2011, 1841–1843). By contrast, more nuanced narratives are likely to produce different identity dynamics, behavioural patterns, and normative outlooks (Hutchison 2016, 276).

The polarized context of war actively encourages such black-and-white interpretations through the social frameworks it provides. As IR scholars note, official responses to trauma often prioritize the protection of the state by shutting down questioning and truncating mourning in order to project strength and restore order in the face of security breaches (Fierke 2004; Edkins 2003; Kinnvall 2013, 154; Lerner 2020). These responses typically rely on simplified narratives and reactive measures that re-establish a sense of state security. By contrast, a genuine working-through process would involve mourning and critically interrogating the political and social arrangements that contributed to the suffering. Such a politically engaged response resists the allure of reductive stories with easily identifiable heroes and villains (Schick 2011, 1837) and may lead to significantly different social outcomes, potentially reducing the likelihood of renewed violence and opening paths towards peace.

The interdisciplinary theoretical insights presented here underscore that interpretations of traumatic events matter: they shape social outcomes and play a critical role in transitions from conflict to peace. Accordingly, this book examines the legacies of war by documenting people's subjective experiences through their personal narratives. It argues that the legacy of war becomes visible in individuals' war stories, their feelings of national attachment, and their visions of the future. Taken together, interpretations of war, senses of national belonging, and imagined futures constitute key dimensions through which the legacy of war is produced and sustained (see Figure 1.1).

These elements of a war legacy are not static; they can be interpreted and reinterpreted by individuals at any time and from new perspectives, with diverse implications for peace. Whenever individuals and communities

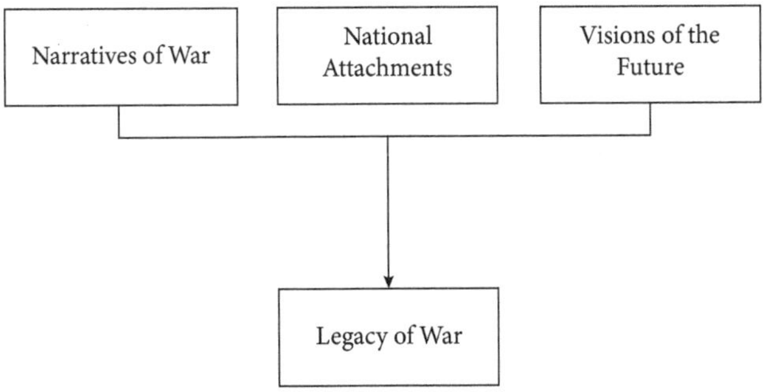

Figure 1.1 Elements Shaping the Legacy of War

work through their war experiences, they actively construct new social realities (Chaitin and Avlagon 2025; Pires, Carvalho, and Ricarte (eds.) 2024). Depending on how war is recounted, framed, and narrated within contemporary societal contexts, these realities evolve. Different interpretations of the same traumatic experience can produce divergent national attachments and normative visions of peace. Hence, multiple modes of engaging with war experiences harbour the potential to generate distinct war legacies and, as a result, multiple possible societal trajectories. Against this backdrop, the crucial question becomes: which subjective interpretations of war trauma are most likely to foster peace rather than renewed conflict in the future?

2

Empirical Encounters

Surveying the Legacies of War in Ukraine, Syria, and Palestine

With the legacy of war defined as the lingering impact that the experience of war leaves on people's social and political attitudes, this book puts individuals who live through wars into focus, particularly their interpretations of the war they have experienced. It views war legacies as manifesting in the 'mind maps' of the people, treating 'each story' as 'a world'. Its goal is to highlight subjective individual voices and the implications these have to create manifold social realities.

The book takes as its case studies three peoples who currently experience war: Ukrainians, Syrians, and Palestinians. They have in common only that they all witness conflict and war. Yet the social context and type of political violence vary widely between them: Ukrainians face an interstate war of outside aggression through Russia; Syrians witness a civil war in the wake of Arab Spring protests to change the regime; and Palestinians live through a protracted, generational conflict with the Israelis about their land. Considering these diverse contexts and types of conflict, the empirical sections of the book present 'cognitive maps' of Ukrainian, Syrian, and Palestinian accounts of *their* war experiences, rather than direct comparisons between them.

To gather the accounts of war from these three peoples, the study employed online surveys among the three groups (Sue and Ritter 2016). The surveys were distributed with the help of research assistants from Ukraine, Syria, and the Palestinian Territories. They were carried out in the native languages, Ukrainian, and Arabic respectively. As the research subjects were in conflict zones or vulnerable because of their displacement, the surveys followed strict ethical guidelines, ensuring their anonymity and informed consent (Matthews and Ross 2010, 308–310). Research on the highly sensitive topic of war and conflict, however, not only entails the responsibility for participants' physical security but also for their emotional experiences. As such, this research was conducted also with an awareness of what Pearlman calls 'emotional ethics' (Pearlman 2023, 1242). Inquiring about war

Legacies of War. Kathrin Bachleitner, Oxford University Press. © Kathrin Bachleitner (2026).
DOI: 10.1093/9780198982913.003.0003

experiences can trigger diverse emotions in participants: on the one hand, it can cause harm through innocuous stimuli; on the other hand, however, talking about painful experiences may also be rewarding. Data that contain people's narratives of their subjective war experiences thus require a high 'emotional sensibility' and the knowledge that 'to enter another's world as a researcher is a privilege [and as such, also a responsibility], not a right' (Fujii 2012, 772).

In total, 287 Ukrainian responses (N=287) were collected in 2023, 337 Syrian responses (N=337) in 2019/20, and 104 Palestinian responses (N=104) in 2020/21. All data were collected in the native languages and translated into English by the research assistants. While sample sizes differ across cases, all analyses are conducted at the individual level, and cross-case comparisons are interpreted cautiously, with attention to contextual specificity rather than statistical inference. The questionnaire followed the three elements of what this book conceptualizes as the legacy of war, prompting respondents to recount in open-ended questions their narratives of the war, their national attachments, and their visions for the future and peace. Notably, this questionnaire reflects an ethnographic sensibility, focusing not only on what people do but also on why they do it. It gleans the meaning that people attribute to their reality (Schatz (ed.) 2009, 5) and seeks to capture people's emotional experiences (Pearlman 2023, 1241).

The Building Blocks of the Survey

The questionnaire consisted of three parts: first, it asked respondents about their narratives of the war they are experiencing; second, it enquired into their national and patriotic feelings towards their country at war, and finally, it probed their conception of the future and visions of peace.

The first set of questions on respondents' war narratives followed what the interdisciplinary literature outlines as the main components of any narrative of war (Alexander's 2012; Bar-Tal 2007): a subjective description of the events that assigns responsibility to a perpetrator and designates victimhood to specific individuals or groups. As a consequence, respondents were asked to determine who, in their view, bears primary responsibility for the war—that is, who the perpetrators are. Wherever responsibility is assigned to a perpetrator, a victim group suffering as a result of the perpetrator's actions is also identified. According to Alexander (2004, 13), it is necessary to provide compelling answers as to which groups of people have been affected by this traumatizing pain. Are particular individuals or groups, or rather 'the people' in general, among those who are suffering the most? Did a singular and delimited group receive the brunt of the pain, or were several groups involved? To find out, survey respondents were asked to specify the party, persons, or groups whom they perceive as the main victims of the war in question.

The second set of questions investigated people's national attachments during wars, and whether they feel more or less proud to be part of their country because of or despite the ongoing conflict. To gauge respondents' levels of national attachment and patriotic pride,[1] the survey followed the World Values Survey's style of questioning[2] by asking respondents to assign themselves to a specific group identity. Within each sample, participants were presented with different answer categories and asked to rank the identities that best described their sense of belonging. Furthermore, levels of patriotic pride were measured by asking repondents to rate, on a scale from 1 to 5, how proud they were to be Ukrainian, Syrian, or Palestinian. After providing this rating ('very high', 'high', 'neutral', 'moderate', or 'low'), respondents were asked about the main reason for their choice.

Finally, the third block of questions gauged respondents' hopes for the future of their country. The survey prompted respondents to describe the preferred borders of the state, the political system, and the ethnic–religious composition of the society. Respondents were also asked whether they would like to return/permanently live in Ukraine/Syria/Palestine.

Case Selection and War Contexts

Ukrainians, Syrians, and Palestinians were selected as this study's sample groups because all three, in diverse ways, are currently living through war. Ukrainians are experiencing a war of external aggression by Russia; Syrians continue to endure a civil war that began with Arab Spring–inspired demands for regime change; and Palestinians are engaged in a generational conflict with Israelis over the same land. The purpose of this study is to map each experience of war separately, without making direct comparisons between the groups. Instead, the empirical case studies aim to delve deeply into individuals' ideas, presenting their voices, interpretations of their war experiences, and hopes for the future.

[1] Please note that the two are often hard to distinguish in empirical data and practice.
[2] World Values Survey. https://www.worldvaluessurvey.org/WVSContents.jsp (viewed 12 July 2025).

The Ukrainian Case: War Background

Since its independence in 1991, Ukraine has been described as a 'fault line' between the East and the West (Howlett 2023a). The country has long been subject to competing interests and ideologies, shaped by a history of imperial rule and occupation. Ukraine was part of the Soviet Union until its dissolution and is today geopolitically positioned between Russia and the European Union (EU). As a result, an antagonism between pro-Russian and pro-European orientations has defined Ukrainian politics and society since independence. The most visible expression of this division surfaced during the Orange Revolution of 2004 and the Euromaidan protests of 2013–2014.

The Orange Revolution was sparked by the 2004 presidential elections, when President Leonid Kuchma selected the former Donetsk governor, Prime Minister Viktor Yanukovych, as his preferred successor. Allegations of widespread electoral fraud triggered mass protests in favour of a Western, democratic, and European orientation. In response, Yanukovych and his team actively promoted the narrative portraying Ukraine—along with the pro-Russian Donbas region— as needing protection from a nationalist Western Ukraine led by the democratic opposition candidate Viktor Yushchenko. During this time, the political map split roughly in half between a predominantly Ukrainian-speaking electorate in the western and central regions, which favoured a path toward European integration, and a largely Russian-speaking population in the country's eastern and southern regions, which preferred to align more closely with the Russia-led Customs Union alongside Russia, Belarus, and Kazakhstan (Haran et al. 2019, 688). In the decade between the Orange Revolution and the Euromaidan protests, the preference for European and Euro-Atlantic integration versus a customs (and in some cases, even military) union with Russia and other Commonwealth of Independent States (CIS) countries constituted the most pronounced cleavage in Ukrainian public opinion.

In this divided context, when President Viktor Yanukovych, then president of Ukraine, decided in November 2013 not to sign the EU–Ukraine Association Agreement and instead pursued closer ties with Russia and the Russia-led Eurasian integration project, a wave of demonstrations erupted in Kyiv's Maidan Nezalezhnosti (Independence Square). Protesters decried governmental corruption, the abuse of executive power, and violations of human rights, calling on Yanukovych to resign. Following the use of violence against protestors, the 'Revolution of Dignity' culminated in February 2014, leading to Yanukovych's ousting. As Russia condemned these events as a coup, pro-Russian protests emerged in southern and eastern Ukraine.

Russia subsequently annexed Crimea, while armed pro-Russian separatists proclaimed the so-called independent 'people's republics' of Donetsk and Luhansk, thereby igniting the Donbas war (Arel and Driscoll 2023; Haran et al. 2019, 689; Wilson 2016). Since then, Russian-supported, if not Russian-led, forces have waged a low-intensity conflict aimed at drawing Ukraine into Russia's sphere of influence and fomenting separatism within the country (Käihkö 2022).

In this altered context, academics across disciplines began to monitor and document the cleavages between the eastern and western regions of Ukraine (Kulyk 2016; Onuch and Sasse 2016). Relying on multiple sources and opinion polls, a broad consensus emerged that these cleavages have been reduced in favour of a more robust Ukrainian national identity. Moreover, studies find that, specifically as a result of Russian interference and aggression, individuals across all regions have increasingly begun to self-identify as Ukrainians and to profess greater pride in being Ukrainian. These developments manifest themselves in stronger attachments to symbols of nationhood, enhanced solidarity with compatriots, and an increased readiness to defend Ukraine. This consolidation of a Ukrainian identity has occurred in tandem with a growing sense of alienation from Russia (Kulyk 2016, 588): 'Contrary to Putin's expectations, the military intervention actually cemented Ukrainian civic and national identity' (Haran et al. 2019, 689). Ukrainians, Onuch and Hale (2022) argue 'identify strongly with their country not because it represents any specific ethnic, linguistic, or religious group but because it represents an inclusive vision of the citizenship as a whole' (Onuch and Hale 2022, 24). They show that the spread of such a 'civic identity' was caused by the coming of age of the 'independence generation' born between 1975 and 1985, but, crucially, also by 'rally around the flag effects' prompted by Russia's actions, and not least by what they term 'the Zelensky effect'. President Zelensky, even before assuming the presidency, embodied and promoted an inclusive Ukrainian citizenship that is explicitly distinct from Russia (Onuch and Hale 2022).

Contrary to Russia's expectations, its military intervention in Ukraine has thus strengthened the Ukrainian civic nation and identity. Poll results have demonstrated that since 2014 there has been a marked trend among the vast majority of Ukrainians indicating that they—including those living in the government-controlled areas of the Donbas—identify first and foremost as Ukrainian citizens (Haran et al. 2019). The Euromaidan Revolution, coupled with mobilization against Russia's aggression, has thus led to profound transformations among Ukrainians. Today, there is a predominant identification with Ukraine as a political state, emphasizing a shared civic identity, rather than ethnic, regional, or local identities. Even in the government-controlled areas of the Donbas, support for Russia-led military or economic integration projects has virtually collapsed, while the number of supporters of Ukraine joining the EU and NATO has increased. As Bekeshikina (2017, quoted in Haran, Yakovlyev, and Zolkina 2019, 692) puts it: 'Putin got Crimea but lost Ukraine'.

These developments became even more apparent following Russia's full-scale invasion on 24 February 2022. In the face of external aggression, a sense of attachment to Ukraine as a nation has remained firmly in place and has indeed grown. This trend is visible first and foremost in the concrete efforts undertaken by the citizenry to support the military and defend Ukrainian territory. Across age groups, Ukrainians have responded by taking up arms or volunteering in a wide-range of capacities. Furthermore, overt expressions of nationhood are evident both domestically and internationally in media platforms and artistic production. In counteracting Russia's aggression, Ukrainian national identity has provided a powerful unifying and motivating force for Ukraine's citizens (Howlett 2023b, 64).[3]

More than two and a half years into the war, the visible legacy of Russia's war on Ukraine is devastating: the invasion has fundamentally disrupted everyday life in Ukraine, resulting in the Russian occupation of Ukrainian territory in the south-east, more than 500,000 estimated military casualties on both sides,[4] at least 10,000 verified civilian deaths, and over 18,500 people injured.[5] In addition to widespread civilian harm and the destruction of infrastructure, more than six million Ukrainians have fled, crossing international borders to neighbouring countries or becoming internally displaced within the comparatively safer western regions of Ukraine.[6]

Timing of the Ukrainian Surveys in the Conflict Cycle: March to May 2023

The online surveys among Ukrainians were carried out from March to May 2023. At this point, the conflict had entered its second year and had largely evolved into a war of attrition, particularly along the Eastern and Southern

[3] See Kyiv International Institute of Sociology, 'Indicators of National-Civic Ukrainian Identity'. 2022. https://www.kiis.com.ua/?lang=eng&cat=reports&id=1131&page=1 (viewed 30 April 2024). See Gallup survey results from 11 October 2023, https://news.gallup.com/poll/512360/ukrainians-future-west.aspx (viewed 30 April 2024).

[4] These numbers were published by the *New York Times* on 18 August 2023, yet numbers are difficult to estimate as Moscow is believed to routinely undercount its war dead, and Kyiv does not disclose official figures. See: https://www.nytimes.com/2023/08/18/us/politics/ukraine-russia-war-casualties.html (viewed 31 October 2025).

[5] See UN Ukraine press release: https://ukraine.un.org/en/253322-civilian-deaths-ukraine-war-top-10000-un-says (viewed 31 October 2025).

[6] See UNHCR's data portal on Ukraine's refugee situation: https://data.unhcr.org/en/situations/ukraine (viewed 31 October 2025).

fronts. This period was generally characterized as a stalemate phase in the conflict cycle, in which neither side achieved decisive breakthroughs, while losses and destruction continued at a significant pace.

One of the most intense battles during this period was the prolonged and highly destructive battle for Bakhmut, a city in the Donetsk region. Russian forces, particularly the Wagner Group, sought to seize the city after months of gruelling combat, while Ukrainian defenders fiercely resisted despite high casualties on both sides.[7] Simultaneously, Russian forces continued a sustained campaign of missile and drone attacks against Ukrainian urban centres and critical infrastructure, exacerbating civilian suffering.[8]

During the survey period, significant preparations were underway for a major Ukrainian counteroffensive. Western allies, including the United States, European Union, and NATO members, were supplying advanced military equipment such as main battle tanks, armoured vehicles, and expanded training for Ukrainian forces. However, the counteroffensive itself did not begin until June 2023, after the conclusion of the survey period.[9]

The humanitarian situation remained severe. By May 2023, approximately 6 million Ukrainians had fled abroad, and an additional 3.7 million were internally displaced.[10] Civilians were coping with war trauma, widespread housing and food insecurity, disrupted healthcare and education services, and ongoing economic instability. However, despite the heavy human and material toll, there was a sense of cautious hope among the population, fuelled by anticipation of the forthcoming counteroffensive and continued international support. At the international level, Ukraine benefited from increasing military, financial, and humanitarian assistance from Western allies. Meanwhile, some Global South actors, including China and Brazil, began to call more publicly for negotiations. Nevertheless, both Ukraine and Russia continued to reject compromises on their core demands.[11]

[7] Institute for the Study of War (ISW). 2023. *Ukraine Conflict Updates, March–May 2023*. Available at: https://understandingwar.org/ (viewed 31 October 2025).

[8] United Nations Office for the Coordination of Humanitarian Affairs (OCHA). 2023. *Ukraine Humanitarian Response Plan 2023*. Available at: https://reliefweb.int/report/ukraine/ukraine-humanitarian-response-plan-2023-february-2023 (viewed 26 April 2025).

[9] International Crisis Group. 2023. *CrisisWatch: Ukraine*. Available at: https://www.crisisgroup.org/europe-central-asia/eastern-europe/ukraine (viewed 26 April 2025).

[10] UNHCR Annual Results Report 2023 Ukraine: https://reporting.unhcr.org/sites/default/files/2024-06/Europe%20-%20Ukraine%20ARR%202023_0.pdf (viewed 24 April 2025).

[11] International Crisis Group. 2023. *CrisisWatch: Ukraine*. Available at: https://www.crisisgroup.org/europe-central-asia/eastern-europe/ukraine (viewed 26 April 2025).

The survey was thus conducted at a time of uncertainty but cautious hope—Ukrainians were aware that a major counteroffensive was coming, but they were still experiencing heavy attacks and losses.

Introducing the Ukrainian Sample Group

The Ukrainian sample consists of 287 respondents. Over one hundred respondents are between twenty and thirty years old, 68% are women, and approximately 80% hold a university degree. These demographic characteristics reflect the recruitment strategy: research assistants, who were university students and young professionals, distributed the online survey through their personal and professional networks, resulting in a convenience-based, non-probability sample. The gender imbalance is also shaped by the broader context of wartime migration. Due to restrictions on men leaving the country, Ukrainians who fled abroad after the invasion are disproportionately women, a pattern reflected in our sample. Figure 2.1 illustrates the details of the sample's characteristics regarding age distribution, gender, and educational level.

The survey was not designed to be representative of the Ukrainian population with respect to age, gender, or education. Instead, the sampling strategy prioritized variation in war exposure, which is central to the study's research aim. To this end, recruitment emphasized geographic diversity and deliberately balanced respondents who remained in Ukraine with refugees residing abroad. Figures 2.2 and 2.3 depict respondents' locations. Figure 2.2 shows the points on the map of Ukraine where respondents were from. Figure 2.3 maps the countries where displaced respondents now reside.

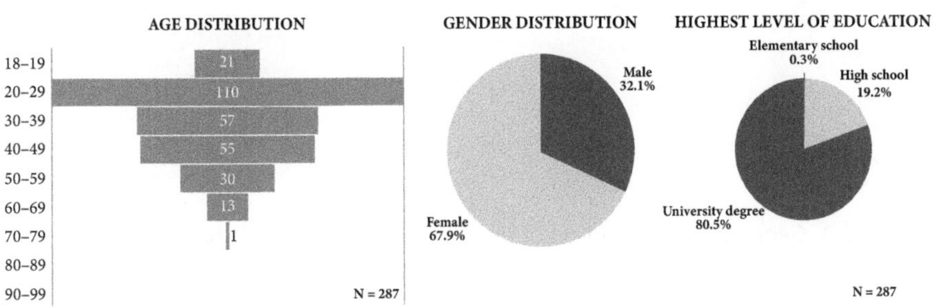

Figure 2.1 Distribution of Age, Gender, and Educational Level in the Ukrainian Sample Group

Figure 2.2 Respondents' Locations within Ukraine at Time of Survey

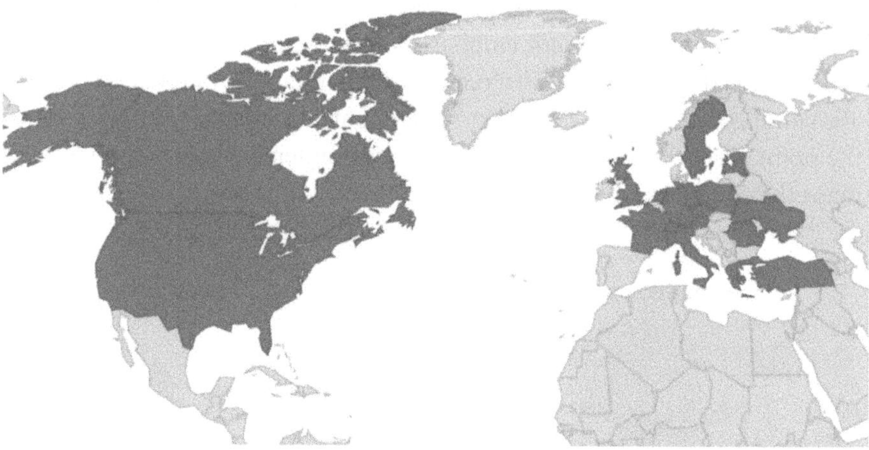

Figure 2.3 Geographic Distribution of Ukrainian Respondents Abroad

As a result, the findings should be interpreted as identifying associations within the sample rather than as providing population-level estimates.

The Syrian Case: War Background

When Hafez al-Assad seized power in Syria after staging a coup in 1970, he took over a 'newly born Republic', one which was 'in many respects a nation without being a nation-state, a political entity without being a political community' (Farouk-Alli 2014, 216). To generate a popular sense of loyalty towards the state—and towards himself—Hafez al-Assad relied heavily on the Baath Party and its widespread appeal to unity among all Arabs. By incorporating a Baathist Arab-nationalist agenda into his regime, he was able to create a unique Syrian state nationalism, one which was united under the banner of *Wahda* (unity), *Huriyya* (freedom), and *Ishtirakiyya* (socialism), or, relatedly, *al-Wahda* (unity), *al-Karama* (honour), and *al-Jaysh* (the army) (Aldoughli 2019; Hinnebusch 2020a; Wedeen 1999). As such, Hafez al-Assad's brand of nationalism mixed elements of Arabism with a socialist ideology. Under the umbrella of an ethnic family of Arabs, sectarian divisions between a Sunni majority population and Assad's affiliation with the Alawite sect were masked. Furthermore, the socialist contract helped to establish a cross-sectarian solidarity between rural Sunnis and Alawites, uniting them behind the Syrian nation (Hinnebusch 1989). Moreover, the Baath Party's Arabism was secular in its orientation. In such a clearly defined secular nation, *ta'ifiyya*, that is, sectarianism, and particularly Islamism, were taboos (Schaebler 2013, 71). The emphasis on secularity, Arabism, and socialism served to promote unity among Syrians and thus to provide legitimacy for Hafez al-Assad's rule (Aldoughli 2022).

However, following the death of Hafez al-Assad in 2000, and the rise to power of his son, Bashar al-Assad, the state-sponsored, secular, Arab, and socialist ties that had united Syrians began to weaken. Towards the end of the twentieth century, Hafez al-Assad's version of state nationalism had all but disintegrated, mainly due to broader regional and national developments. First, across the Middle East, Arabism had rapidly declined in popularity and was slowly replaced by Political Islam as an ideological orientation (Dawisha 2000). Once Political Islam had acquired strength, sectarian divisions and competing nationalistic tendencies began to tear apart the 'Arab family'. As a consequence, religious and sectarian ideas that had once persisted tacitly under the umbrella of the officially promoted Arab identity resurfaced (Ajami 1992; Browers 2009; Olmert 2011).

Second, as the Pan-Arab idea continued to lose its appeal on the regional level, so did the socialist contract with rural populations in Syria (Ziser 2001). Due to economic recession and newly introduced liberal economic reforms, Bashar al-Assad's ability to co-opt the support of rural constituencies through economic benefits was much more limited than that of his father. In fact, the economy's shift towards neoliberalism gradually required him to abandon his father's socialist contract with the Syrian people—yet it was precisely this contract that had previously secured their loyalty (Hinnebusch 2020b; Perthes 2004).

In response to this changed context, Bashar al-Assad refurbished the fading identity of the nation by increasingly conflating it with the identity of the president (Ismail 2011, 542). He refashioned his father's Arab and socialist strategies around his own leadership and personal authority (Wedeen 1999). As such, he effectively 'reimagined' the country as *Suriyya al-Assad*, making himself the linchpin of Syrian national identity. Syria now went from being a country for all 'Syrian Arabs' to becoming 'Assad's country', a country open to all those who follow the president (Chevée 2022). From father to son, Syria's uniting social purpose therefore shifted from socialism to capitalism, and from a national belonging that was inclusive towards Syrian Arabs to one that hinged on an individual's loyalty to the Syrian regime. It was within this context of national-ideological transformation that the Arab Spring began in 2011.

What had started as popular uprisings against the ruling Assad regime in 2011 quickly escalated into a brutal civil war between government forces and a fragmented coalition of opposition groups. The conflict soon became increasingly complex as a wide range of armed actors emerged, including secular and Islamist opposition factions, Kurdish forces in northern and eastern Syria, and transnational jihadist groups such as the Islamic State (IS). Over time, regional and international powers intervened directly or indirectly, transforming the conflict into a multi-layered war. Iran and Hezbollah provided military and logistical support to the Assad regime, while Russia intervened militarily in 2015, significantly altering the balance of power in the regime's favour. Turkey supported various opposition factions and later deployed forces in northern Syria, while the United States and allied states intervened primarily through air operations and support for Kurdish-led forces against IS (Phillips 2017). These overlapping domestic, regional, and international interventions prolonged the conflict and contributed to its transformation into one of the most complex and destructive wars of the early twenty-first century.

The traumatic consequences inflicted on the Syrian people and society as a whole are immense: entire cities and villages swept away; an estimated 306,000 civilians were killed[12] and 1.9 million wounded (Phillips 2017, 1); the war also triggered one of the largest refugee waves in human history—roughly

[12] Numbers based on UN Human Rights Office estimates in June 2022. See: https://www.ohchr.org/en/press-releases/2022/06/un-human-rights-office-estimates-more-306000-civilians-were-killed-over-10 (viewed 7 December 2025).

6 million Syrians have fled the country to neighbouring Middle Eastern states and beyond. According to the United Nations High Commissioner for Refugees (UNHCR), a further 7.4 million Syrians were internally displaced (IDPs).[13]

In late 2024, the Syrian conflict entered a new stage following a rapid military offensive by opposition forces led by Hay'at Tahrir al-Sham (HTS) and allied factions, which resulted in the collapse of regime control across large parts of the country. By December 2024, Bashar al-Assad had fled to Russia, and more than five decades of Assad-family rule came to an end. HTS, under the leadership of Ahmed al-Sharaa, subsequently assumed de facto authority over the capital and a number of major urban centres, announced an interim governing arrangement, and initiated a transitional political process, including plans for new constitutional and administrative frameworks. Despite the change in power, Syria remained territorially fragmented, security conditions varied across regions, and incidents of violence continued. The humanitarian situation also remained severe: millions of Syrians continued to be internally displaced or to live as refugees abroad, and although some return movements were reported following the fall of the regime, these remained uneven and constrained by ongoing insecurity, widespread destruction, and a lack of basic services.[14]

Timing of the Syrian Surveys in the Conflict Cycle: May to June 2019 and July to October 2020

Online surveys were conducted among Syrians in two cycles: the first between May and June 2019, and the second from July to October 2020. These surveys captured Syrian narratives at two distinct, but equally critical, stages in the protracted civil war, then entering its ninth and tenth years. By this time, the conflict landscape was highly fragmented, with territorial control divided among several actors: the Syrian government backed by Russia and Iran; opposition groups, including Turkish-supported factions; Kurdish-led Syrian Democratic Forces (SDF) backed by the United States; and jihadist groups such as Hay'at Tahrir al-Sham (HTS) dominating Idlib.[15]

[13] The (updated) numbers on Syrian refugees and IDPs can be checked at UNHCR, 'Refugee Data Finder: https://www.unhcr.org/refugee-statistics (viewed 7 December 2025).

[14] See UNHCR, 'Syrian Arab Republic country data': https://www.unhcr.org/where-we-work/countries/syrian-arab-republic (viewed 7 December 2025) and UNHCR, 'Syria Regional Flash Update': https://www.unhcr.org/syria-emergency(viewed 7 December 2025).

[15] United Nations Security Council. 2019. *Report on the Situation in Syria (S/2019/541)*. Available at: https://undocs.org/S/2019/541 (viewed 26 April 2025).

The first survey cycle (May–June 2019) coincided with the escalation of the Idlib crisis, when the Assad regime, with Russian air support, launched a major offensive to retake Idlib Province, the last significant rebel stronghold. The offensive, beginning in late April 2019, resulted in some of the deadliest violence against civilians since 2017.[16] Intense airstrikes and shelling across Idlib and northern Hama displaced hundreds of thousands towards the Turkish border by mid-2019.[17] At the same time, those Syrians who already lived as refugees abroad, particularly in the neighbouring countries of Turkey, Lebanon, and Jordan, had to deal with a growing pressure to return, even though the security situation in Syria made return nearly impossible. War trauma, fear, and a high level of scepticism about any prospect for peace dominated the general mood of Syrians at the time.[18]

The second survey cycle (July–October 2020) took place during a relatively quieter, though still unstable, period following a March 2020 ceasefire in Idlib, brokered by Russia and Turkey. However, the ceasefire was tenuous, with frontlines remaining active and sporadic clashes continuing.[19] The political-military landscape had shifted significantly after Turkey's October 2019 incursion into north-east Syria following the partial withdrawal of US forces and resulting in new alliances between the Kurdish SDF, the Syrian government, and Russia.

The humanitarian situation in 2020 was dire, and by the time of the second survey round, Syria counted over 6 million IDPs, and more than 5.6 million refugees in neighbouring countries.[20] In 2020, the Covid-19 pandemic further strained Syria's already fragile healthcare system,[21] while an accelerating economic collapse compounded the population's hardships. By June/July 2020, the Syrian pound had reached record lows, triggering severe inflation, widespread food insecurity, and a humanitarian

[16] United Nations Office for the Coordination of Humanitarian Affairs (OCHA). 2019. *North-West Syria Situation Report No. 10 (28 June 2019)*. Available at: https://reliefweb.int/report/syrian-arab-republic/syrian-arab-republic-north-west-syria-situation-report-no-10-28-june-2019 (viewed 26 April 2025).

[17] United Nations High Commissioner for Refugees (UNHCR). 2019. *Thousands Displaced in Northwest Syria*. Available at: https://www.unhcr.org/news/briefing/2019/6/5d13b5c04/thousands-displaced-north-west-syria-violence-unhcr-appeals-protection.html (viewed 26 April 2025).

[18] Human Rights Watch. 2019. *Lebanon: Summarily Deporting Syrians*. Available at: https://www.hrw.org/news/2019/10/24/lebanon-summarily-deporting-syrians (viewed 26 April 2025).

[19] International Crisis Group. 2020. *The Best of Bad Options for Syria's Idlib*. Available at: https://www.crisisgroup.org/middle-east-north-africa/eastern-mediterranean/syria (viewed 26 April 2025).

[20] United Nations High Commissioner for Refugees (UNHCR). 2020. *Syria Emergency Overview*. Available at: https://www.unhcr.org/syria-emergency.html (viewed 26 April 2025).

[21] World Health Organization (WHO). 2020. *COVID-19 Situation Reports—Syria*. Available at: https://www.who.int/emergencies/situations/covid-19-syria-situation-reports (viewed 26 April 2025).

emergency.[22] Poverty became pervasive, basic goods such as bread were unaffordable for most households, and many communities—particularly in non-government-controlled areas—were almost entirely dependent on humanitarian aid. However, aid access was heavily politicized and subject to growing restrictions.[23]

The Syrian respondents thus are telling their war narratives while living through fresh displacement waves because of renewed violence in Idlib and north-eastern Syria, a fragmented and shifting political and military landscape, a collapsing economy and healthcare system, exacerbated by the Covid-19 pandemic, and deep uncertainty about the future and their return.

Introducing the Syrian Sample Group

In the Syrian sample of 337 participants, over one hundred are between the ages of thirty and forty; 79% are men, and 74% hold a university degree. These demographic characteristics reflect the recruitment strategy: research assistants, who were young professionals, distributed the online survey through their personal and professional networks, resulting in a convenience-based, non-probability sample. The gender imbalance in the sample reflects broader patterns of visibility and mobility in the Syrian context. Men tend to be more visible in the public sphere, and Syrian migrants abroad are disproportionately male, both of which are reflected in the composition of the sample. Figure 2.4 presents detailed characteristics of the sample in terms of age distribution, gender, and education level.

As in the other samples, the survey was not designed to be representative of the Syrian population with respect to age, gender, or education. Instead, the sampling strategy prioritized geographic diversity and variation in war-related experiences, which are central to the study's research objectives. Figure 2.5 shows respondents' locations within Syria at the time of survey completion, while Figure 2.6 depicts the countries to which displaced respondents had fled.

[22] United Nations Office for the Coordination of Humanitarian Affairs (OCHA). 2021. *Humanitarian Needs Overview: Syrian Arab Republic (March 2021)*. Available at: https://reliefweb.int/report/syrian-arab-republic/2021-humanitarian-needs-overview-syrian-arab-republic-march-2021 (viewed 26 April 2025).

[23] United Nations Office for the Coordination of Humanitarian Affairs (OCHA). 2020. *Cross-Border Humanitarian Update (1 June 2020)*. Available at: https://reliefweb.int/report/syrian-arab-republic/syria-cross-border-humanitarian-update-1-june-2020 (viewed 26 April 2025).

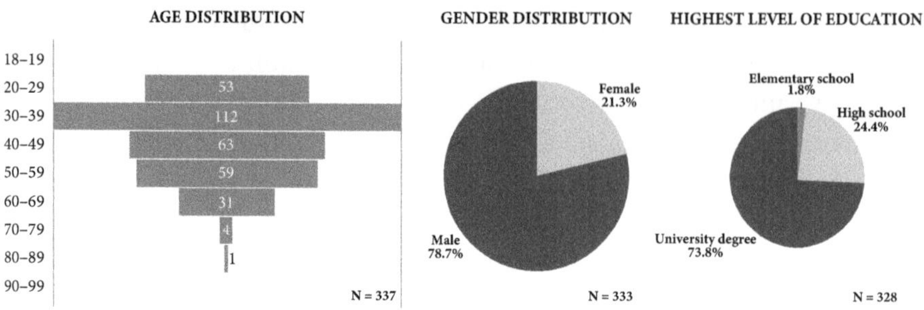

Figure 2.4 Distribution of Age, Gender, and Educational Level in the Syrian Sample Group

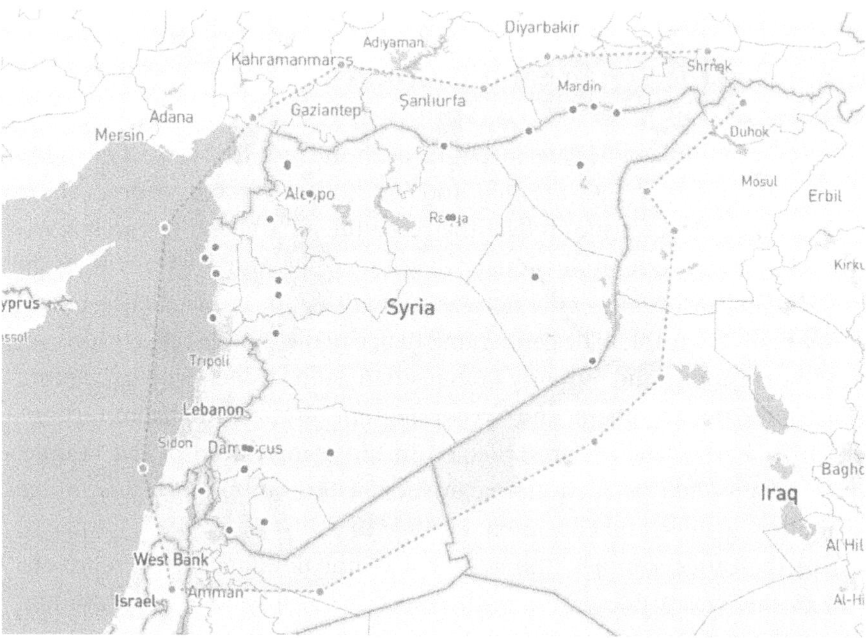

Figure 2.5 Respondents' Locations within Syria at Time of Survey

The Palestinian Case: War Background

The conflict between Israelis and Palestinians dates back more than a century (Tessler 2009). At its core, it is a struggle between two national movements—Zionism and Palestinian nationalism—both aspiring to claim the same stretch of land. While the ongoing cycles of violence have become entangled with ethnicity, religion, political ideology, and global power structures, the conflict remains, at its essence, 'a modern territorial contest: two nations, one land'.

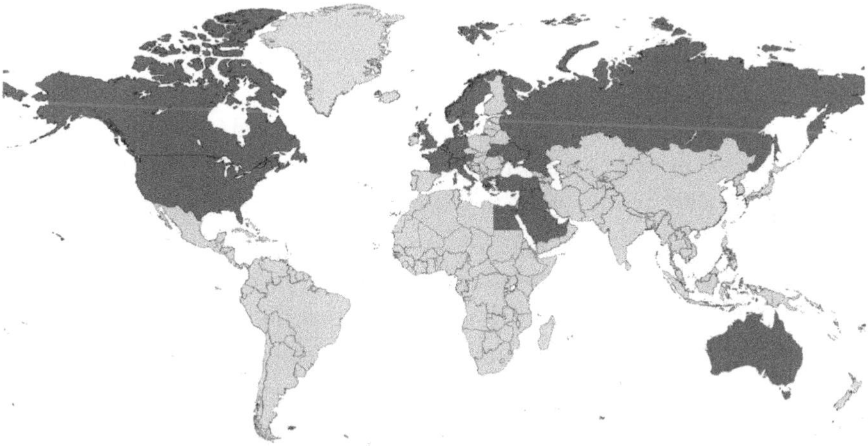

Figure 2.6 Geographic Distribution of Syrian Respondents Abroad

As Bunton (2013, xii) puts it, 'The main challenge to resolving the conflict is essentially one of drawing borders.'

The historical roots of the conflict stem from the late nineteenth and early twentieth centuries, when Jewish immigration increased in response to European anti-Semitism. Supported by an ideological infrastructure aimed at the creation of a Jewish state (see specifically Theodor Herzl's book *Der Judenstaat*), inter-group tensions between the Jewish immigrants and Arab residents soon escalated into violence. While the British who held the Mandate over Palestine struggled to balance the interests of both groups, the Second World War and the ensuing Holocaust had rendered the Zionist demand for a Jewish state urgent in the eyes of the international community. As a result, in 1947, the United Nations adopted Resolution 181, known as the Partition Plan, dividing the British Mandate of Palestine into Arab and Jewish states. While the State of Israel was proclaimed, accompanied by widespread enthusiasm and celebrations, the Palestinian Arabs did not accept the Partition Plan, and the first Arab–Israeli War began.

In Israel, this war became known as 'the War of Independence', whereas the Palestinians termed it '*al-Nakba*' ('the catastrophe'). By the war's end in 1949, approximately 750,000 Palestinians were displaced, while the territory of the State of Israel increased. As such, in Israel's national memory, this war is hailed as a liberation, with Israelis celebrating a heroic end to centuries of suffering and powerlessness in the diaspora. In contrast, the Palestinian Arab society was largely destroyed, and its population was forcibly displaced throughout the region. In Palestinian national memory, the '*Nakba*' became synonymous with their dispossession and expulsion from the land of their ancestors (Bunton 2013, 55).

The protracted, generational conflict that followed entrenched contradictory views on each other's trauma and grievances (Rotberg (ed.) 2006). Through the UN Partition plan, the tragic legacy of the Holocaust had become entangled with the Palestinian *Nakba*, creating opposing narratives of conflict. The Israeli national narrative is one of Zionist acceptance and Arab rejection. The mainstream Israeli view holds that the Arab side initiated the war and therefore bears responsibility for its consequences, including the displacement of Palestinian refugees who are often described as having left their homes 'voluntarily' following calls from Arab governments (Shlaim 1995). From the Palestinian perspective, responsibility for 1948 and the *Nakba* lies with the superpowers, especially Great Britain for installing a Jewish state on Arab lands, while Israel is seen as fully responsible for the expulsion of Palestinians from their homes (Jawad 2006, 73–74).

In the following decades, Israel focused on consolidating its economy, absorbing new waves of Jewish immigrants, and maintaining its defences against its Arab neighbours. With the outbreak of the Six-Day War in 1967, however, another major turning point occured, with repercussions that persist to this day. This war—which only lasted six days—resulted in a decisive Israeli victory. As a consequence, Israel's power, inner morale, and external image shifted from a weak and defensive state to a strong, offensive, and occupying force in the Middle East. The territory under Israeli control tripled, and approximately 300,000 more Palestinians were displaced (Oren 2002). Israel captured the Sinai and Gaza from Egypt, the West Bank, including East Jerusalem from Jordan, and the Golan Heights from Syria. In contrast to the Israeli victory stood the Arab defeat. The Pan-Arab ideology collapsed as a result of the failure to jointly 'liberate Palestine'. In this context, a process of introspection began that led to a national reorientation of Arab countries and the rise of Political Islam as an increasingly influential ideology. As a consequence of these developments, the Arab–Israeli conflict transformed into a Palestinian–Israeli conflict, with religion assuming greater salience (Ajami 1978).

First, the national reorientation of Arab countries propelled the Palestine Liberation Organization (PLO) into a more central role, with Palestinians under Yasser Arafat's leadership increasingly taking political and military matters into their own hands. Egypt, previously the strongest supporter of the Palestinian cause, underwent a period of national reorientation—accelerated by the leadership transition from Gamal Abdel Nasser to Anwar Sadat. These domestic political changes in Egypt culminated in a bilateral peace agreement with Israel in 1979, through which Egypt regained the Sinai Peninsula. This agreement represented a milestone in Israel's post-1967 strategy and followed the logic of 'land for peace'. However, this logic did not extend to the occupied territories of the West Bank and Jerusalem. These locations, which hold profound religious significance for Jews and Muslims alike, heightened the centrality of religion to the conflict on both sides. On the Palestinian side, Islamist movements became increasingly associated with the armed struggle to liberate Palestine. On the Israeli side, the settler movement emerged. Driven by religious, political, nationalistic, and economic motivations, it continues to represent a significant obstacle to the exchange of 'land for peace' with the Palestinians (Gorenberg 2006). Viewed through religious and ideological lenses, the Holy Land is simply not open to negotiation.

Yet, the two sides once came close to finding a compromise in the so-called Oslo Accords in 1993. This was the first time the two parties mutually acknowledged one another's existence and agreed on a roadmap toward peace by way of a two-state solution. The borders were set at the pre-1967 lines, encompassing the West Bank with East Jerusalem and the Gaza Strip as the territory of a future Palestinian state. The peace process, however, was soon derailed by periods of violence and unrest, political assassinations, the expansion of Jewish settlements, a persisting Israeli military presence, the erection of a 'security barrier' that restricted the movement of people and goods, and finally, internal divisions among the Palestinians that culminated in the split between Fatah, the party that negotiated the Oslo Accords, and the Islamist Hamas, which opposed the peace plan.

Since Hamas's electoral victory in 2006, the group has controlled Gaza, while Fatah has continued to govern the West Bank. The two parties have remained divided in their political, ideological, and strategic outlooks on peace. Moreover, from Gaza, Hamas has persisted in its armed struggle against the Israeli occupation. The repeated launching of rockets has prompted four military operations, further deepending distrust and grievances on both sides. To make matters worse, on 7 October 2023, Hamas-led militant groups launched a surprise attack on Israel, killing more than one thousand civilians and taking more than two hundred hostages. The attack sparked the fifth Gaza war since 2008, which, at the time of writing, was ongoing and had already been described as the deadliest war for Palestinians ever fought in the history of the conflict.[24]

[24] See the data from the United Nations Office of the Coordination of Humanitarian Affairs, at: https://www.ochaopt.org/data/casualties (viewed 11 September 2024).

In the face of this protracted, decades-long conflict between Israelis and Palestinians, parties and various players on each side still disagree on the key questions of borders, Jerusalem, and settlements (Shlaim 2014). Furthermore, the Palestinian refugee issue remains unresolved to this day. The conflict has been aggravated by significant power asymmetries between Israelis and Palestinians in economic development, military strength, and international support, all of which have been further exacerbated by Israel's close alliance with the United States. It is within this context of unresolved territorial, political, and identity-based disputes that political attitudes toward violence, compromise, and coexistence are shaped.

Timing of the Palestinian Surveys in the Conflict Cycle: November 2020 to April 2021

Palestinians respondents filled in the survey between November 2020 and April 2021. This time frame coincided with mounting instability and an increasing probability of escalation and large-scale violence, even though major hostilities had not yet resumed.

The year leading up to November 2020 was marked by significant political developments that reshaped the landscape of the conflict between Israel and the Palestinians. In January 2020, the Trump administration unveiled the 'Peace to Prosperity' plan, widely interpreted as institutionalizing asymmetries by endorsing Israeli sovereignty over large parts of the West Bank, while offering Palestinians limited autonomy under restrictive conditions. This initiative was overwhelmingly rejected by Palestinian political actors and civil society as legitimizing the de facto annexation of occupied territories.[25] Simultaneously, the Covid-19 pandemic exacerbated existing humanitarian crises, particularly in Gaza, and further restricted movement, economic activity, and political mobilization across the West Bank and East Jerusalem.[26]

Importantly, the lived experiences of respondents diverged significantly between Gaza and the West Bank: respondents in Gaza endured conditions of protracted humanitarian crisis, underpinned by an enduring

[25] European Parliamentary Research Service. 2020. *Peace to Prosperity: A Political Overview.* Brussels: European Union. Available at: https://www.europarl.europa.eu/EPRS/EPRS-AaG-646166-US-peace-plan-Middle-East-FINAL.pdf (viewed 25 April 2025).

[26] United Nations Office for the Coordination of Humanitarian Affairs (OCHA). 2021. *Occupied Palestinian Territory: Humanitarian Response Plan 2021.* Available at: https://www.un.org/unispal/wp-content/uploads/2020/12/OCHAHUMRESPLAN_151220.pdf (viewed 26 April 2025). United Nations Relief and Works Agency for Palestine Refugees in the Near East (UNRWA). 2020. *Annual Operational Report 2020.* Available at: https://www.un.org/unispal/wp-content/uploads/2021/08/UNRWAANNUALTPT_100821.pdf (viewed 26 April 2025).

Israeli–Egyptian blockade, recurrent episodes of low-intensity armed clashes, and systemic economic deprivation. The situation in Gaza was particularly acute, with persistent border tensions, high unemployment, electricity shortages, and an acute sense of vulnerability to renewed large-scale conflict.[27]

Respondents in the West Bank, in contrast, faced an escalation of political repression, settler violence, settlement expansion, and periodic confrontations with Israeli security forces, especially in and around East Jerusalem. Movement restrictions, house demolitions, and settler-related violence contributed to an increasingly coercive environment.[28]

Shortly after the survey concluded, in May 2021, Israel launched *Operation Guardian of the Walls*, which resulted in widespread destruction, particularly in Gaza.[29] This operation was precipitated by a series of flashpoints in East Jerusalem, notably the pending evictions of Palestinian families in Sheikh Jarrah and Israeli police actions at the Al-Aqsa Mosque compound. The survey thus captured Palestinians' experiences and narratives at a time when structural injustice and political conflict were converging toward open violence. Furthermore, the general sentiment was likely one of popular disillusionment with the prospects for a negotiated settlement and a sense of abandonment by the international community, particularly in light of Trump's policies.

Introducing the Palestinian Sample Group

The Palestinian sample consists of 104 respondents. The majority of respondents hold a university degree (approximately 90%), while the age and gender distribution is more balanced than in the other samples: 45% of respondents are in their twenties, 31% in their thirties, and 58% are men. These demographic characteristics reflect the recruitment strategy of the study. Research assistants, who were university students and young professionals, distributed the online survey through their personal networks,

[27] United Nations Relief and Works Agency for Palestine Refugees in the Near East (UNRWA). 2021. *Emergency Appeal 2021*. Available at: https://www.un.org/unispal/wp-content/uploads/2021/03/UNRWAEMERAPPEAL_020321.pdf (viewed 26 April 2025).

[28] United Nations Office for the Coordination of Humanitarian Affairs (OCHA). 2020. *Occupied Palestinian Territory: Humanitarian Needs Overview 2020*. Available at: https://palestine.un.org/sites/default/files/2021-12/country_humanitarian_report_2020_0.pdf (viewed 26 April 2025).

United Nations Economic and Social Council. 2021. *Economic and Social Repercussions of the Israeli Occupation on the Living Conditions of the Palestinian People in the Occupied Palestinian Territory, including East Jerusalem, and the Arab Population in the Occupied Syrian Golan* (A/76/94–E/2021/73). Available at: https://documents.un.org/doc/undoc/gen/n21/166/61/pdf/n2116661.pdf (viewed 26 April 2025).

[29] United Nations Office for the Coordination of Humanitarian Affairs (OCHA). 2021. *Occupied Palestinian Territory: Humanitarian Response Plan 2021*. https://www.un.org/unispal/wp-content/uploads/2020/12/OCHAHUMRESPLAN_151220.pdf (viewed 26 April 2025).

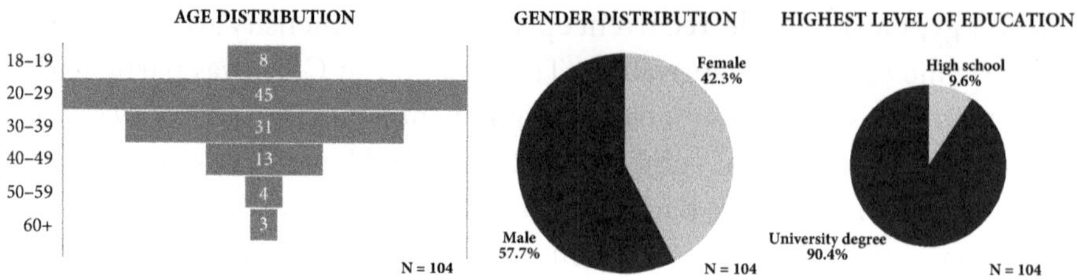

Figure 2.7 Distribution of Age, Gender, and Educational Level in the Palestinian Sample Group

Figure 2.8 Respondents' Locations within the Palestinian Territories/Israel at Time of Survey

resulting in a convenience-based, non-probability sample. Figure 2.7 presents detailed information on respondents' age distribution, gender, and educational attainment.

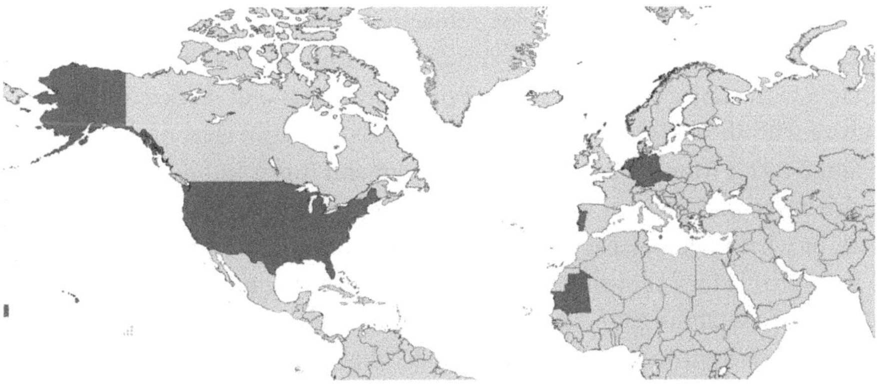

Figure 2.9 Geographic Distribution of Palestinian Respondents Abroad

Consistent with the overall research design, the survey did not aim to achieve demographic representativeness by age, gender, or education. Instead, the sampling strategy prioritized variation in conflict exposure and geographic coverage, which are central to the study's objectives. Accordingly, recruitment focused on reaching respondents residing in Gaza, the West Bank, Israel, and the Palestinian diaspora abroad. At the time of the survey, 82% of respondents were living in the Palestinian Territories, either in Gaza or the West Bank, while a smaller number resided in Israel (see Figure 2.8). The remaining 18% lived abroad, primarily in Europe and the United States (see Figure 2.9).

Empirical Challenges

This study relied on "virtual cognitive maps" extracted from online survey data, a strategy that helped mitigate access constraints in war zones by leveraging modern digital technologies. Online surveys are particularly well suited to conflict settings because of their mobility, accessibility, and the anonymity they afford respondents (Sue and Ritter 2016). Moreover, most of the data were collected during the COVID-19 pandemic, when in-person research was largely infeasible. Nevertheless, this approach also entailed methodological risks and introduced potential sources of bias.

Most dangerous for this book's research aims are the pitfalls of an ideological bias. Researching war through the eyes of the affected people bears a high risk of inciting partiality on the part of the researcher and readers. People who witness wars usually are convinced that they 'have lived the deepest truth of the war' (quotation from an interviewee in Los Horcones, 1992, in: Wood 2012, 31). Using people's subjective stories as data thus can easily sway researchers and readers into certain emotions, risking the superseding of facts with subjective judgements. Both researchers and readers can sympathize with the victims and adopt the polarized negative view of opponents (Käihkö 2022; Monroe 2011, 323). However, and this is important, the aim of this research is not to invite judgement on which narrative is closest to facts, or which perspective is deserving of support and sympathy, but rather to show that actors' subjective self-understandings and interpretations of the war provide vital insight for social explanation.

It is precisely in response to these risks that the study adopts a micro-foundational, interpretive analytical strategy. Respondents' accounts serve as the empirical basis for exploring social outcomes, with particular attention to how interpretations of war reflect political attitudes and orientations (Kurzman 2004). Of course, our emotions are not neutral in the face of war-torn societies and people's accounts of suffering. Yet, the many original extracts from the data in this book serve first and foremost to illuminate their potential for understanding social outcomes rather than fostering moral support, judgement, or solidarity. If anything, the quotations may encourage readers to listen closely to different people's voices while analytically recognizing the power that diverse individual interpretations have on the prospects for peace.

Employing people's subjective interpretations of war as data can not only spark partiality, but such data also contain a large degree of unreliability (Matthews and Ross 2010, 271–273). Personal narratives always include omissions, and memories carry deliberate or inadvertent misrepresentations (Pearlman 2016; 22). Moreover, the political stakes of war mean that narrators are highly emotional, and often respondents seek to highlight or censure information according to the war aims. Additionally, respondents in surveys can easily claim lofty values and motives that follow the group's war aims without real-world repercussions. Furthermore, their ideas are often not their own, but rather each community has its hardened social scripts that reinforce similar patterns of argument and discourse (Pearlman 2017, 503). To add another layer of caveat, the responses reflect not only how individuals chose to present themselves, but also how they present themselves to a research project that originates at the University of Oxford. Furthermore, responses depend on the time, context, and mood in which respondents find themselves, and may vary significantly when any of these factors change (Bachleitner et al. 2010). Rather than constituting a limitation, this variability is central to the analytical value of the data, as the study treats subjective narratives as potential sources of social explanation. The research therefore does not seek to adjudicate which narratives are closest to objective truth, but to demonstrate how diverse interpretations of war lead to the formation of different legacies of war.

Finally, the online survey method employed in this study entails well-known risks of selection bias (Matthews and Ross 2010, 303; Sue and Ritter 2016). Recruitment proceeded in two stages. First, research assistants developed a sampling frame intended to reach respondents across different geographic locations, including within countries of origin and across refugee-hosting contexts abroad. Second, survey links were distributed using a snowball approach through respondents' social and professional networks. As a consequence, participation was more likely among individuals with access to technology and higher levels of education.

These recruitment dynamics shape the composition of the samples. The Syrian sample includes a higher proportion of middle-aged men than would be expected in the broader population. The Ukrainian sample contains a larger share of young women, particularly among refugee respondents, reflecting restrictions on men's ability to leave the country. The Palestinian sample is smaller than the other cases due to comparatively low response rates. While these characteristics limit demographic representativeness, representativeness is not central to the aims of this study. Rather than projecting beliefs onto entire populations, the analysis seeks to foreground respondents' voices and to examine how their subjective interpretations of war illuminate social and political outcomes.

Research Methods and the Presentation of Results

This study investigates the legacies of war as they take shape in the minds and narratives of individuals living through ongoing conflict. To trace how these legacies are formed—through personal accounts of wartime experience, national identification, and imaginaries of peace—a mixed-methods approach was employed. The methodological design was developed to remain attentive both to individual voices and to broader patterns in the samples, enabling a layered analysis of how narratives contribute to the construction of political and social visions for the future.

From the outset, the analysis maintained continuous and iterative engagement with the collected data. While the core aim was to allow respondents to speak in their own terms about war and its meaning, the sheer volume and diversity of individual narratives necessitated systematic synthesis to identify broader patterns and make meaningful connections between empirical data and theoretical insights. To honour the richness of individual accounts while producing analytically coherent findings, the study combined qualitative and quantitative techniques in a reflexive, three-stage process.

The first stage involved a thematic analysis of the raw textual data, which served as a crucial interpretive step for identifying emergent themes and recurring ideas. This step enabled the identification of key themes, narrative structures, and interpretive logics related to the three focal areas of the study: personal accounts of conflict events and actors, expressions of national attachment, and visions of the future. Thematic coding thus created an analytical bridge between rich textual data and subsequent quantitative processing (Matthews and Ross 2010, 373–375).

In the second stage, a statistical analysis of the thematic codes allowed for descriptive mapping of trends in the three case groups—Ukrainians, Syrians, and Palestinians. This quantitative layer supported the identification of broader patterns in narrative content and sentiment. However, as quantitative aggregation can obscure the complexity and nuance of individual meaning-making (McNabb 2021, 361–373), a third analytical step returned to a close, qualitative engagement with the data that re-examined selected responses in depth, with the goal of recentring the voices behind the patterns.

This final stage employed narrative analysis to explore not merely the content of the responses, but the form and structure of the stories that participants told. Here, the focus shifted from what was said to how it was said: the structure, sequencing, and discursive strategies used by respondents to make sense of war, narrate their national belonging, and imagine the future. Particular attention was paid to the narrative construction itself—the sequencing of events, the use of explanatory language, and the rhetorical strategies deployed to convey personal and collective meaning. This method enabled an understanding of narratives as performative acts that reflect and shape collective identity and political possibilities. Close attention was given to the ways in which narratives were informed by their temporal and political contexts, and to how individuals framed their stories amid the pressures, uncertainties, and shifting meanings of ongoing conflict (Matthews and Ross 2010, 388–389).

Together, this threefold methodological approach—moving from thematic interpretation to statistical mapping and back to narrative depth—was designed to illuminate how war is rationalized, how community is interpreted, and how the future is reimagined by those who live through it. In doing so, the book offers a framework for analysing the discursive foundations of war's enduring legacy in society.

The results from these analyses are presented in the three empirical chapters of the book. Chapter 3 focuses on Ukrainian, Syrian, and Palestinian *narratives of war*. Chapter 4 discusses people's *national attachments during war*. Chapter 5 explores the *visions of the future* that emerge from these samples.

Chapter 3 describes people's narratives of war, allowing respondents from the three groups to share their stories and perspectives on the conflicts they are experiencing. To aggregate the diverse responses, narratives were categorized based on whom respondents identified as the main victims and perpetrators of political violence. They were further coded as either 'political' or 'ethnic/sectarian' interpretations of the conflict.

Chapter 4 analyses people's national attachments in times of war—that is, their feelings towards their country. Respondents first ranked their patriotic pride numerically, and then, in a second step, explained why they do or do not feel proud to be part of their national group. These open-ended responses were coded into categories reflecting 'political' or 'cultural' reasons for patriotic pride.

Chapter 5 explores people's visions of peace and the future of their country. It lays out how respondents from each sample group understand their ideal state's borders, political system, and religious and ethnic composition. These visions were categorized according to whether they reflect a 'civic' or an 'ethnic/sectarian' idea of the country's future.

3
Narratives of War

'Conflicts depend on narratives and cannot exist without an explanation of how and why the battles began, and why one side, and only one side, is in the right' (Rotberg 2006, Preface). In practice, this means that war accounts typically follow a narrative arc that assigns roles of perpetrator and victim, structuring perceptions of responsibility and suffering. Such representations are inherently subjective and emotionally charged, as 'every conflict is justified by a narrative of grievance, accusation, and indignity' (Rotberg 2006, Preface).

Chapter 3 examines the narratives of war articulated by Ukrainians, Syrians, and Palestinians who shared their perspectives on political violence through online surveys. When individuals attempt to make sense of war as they experience it, they often divide the warring parties into two opposing moral camps—victims and perpetrators, good and evil. This moral bifurcation aligns with what Bar-Tal (2007) refers to as the 'ethos of conflict', within which individuals draw sharp boundaries between in-groups and out-groups, with the in-group becoming morally elevated through its victimization and belief in the justness of its cause (Noor et al. 2017; Vollhardt and Bilewicz 2013). These beliefs are usually accompanied by the delegitimization, and in some cases, dehumanization of the opponent (Bar-Tal and Hammack 2012).

Moreover, as the theories have suggested, such subjective framings develop within the social reality people experience, and the dominant public discourse put forward by politicians and media at the time. Especially Jeffrey Alexander's (2012, 17–18) 'social trauma' concept highlights the social dimension in all components of war narratives: in people's description of the traumatizing event; in their assignment of responsibility and suffering to specific individuals or groups; and in how they formulate an explanation that connects the perpetrators to their victims (Alexander 2012, 37). As scholars of International Relations have noted, wartime polarization often leads political leaders to adopt and disseminate official narratives that frame conflicts in starkly moralistic and binary terms. Such framings serve strategic purposes: in the face of acute security threats, political elites frequently seek to shut down questioning and dissent and curtail public mourning in order to project

Legacies of War. Kathrin Bachleitner, Oxford University Press. © Kathrin Bachleitner (2026).
DOI: 10.1093/9780198982913.003.0004

resilience, assert control, and preserve an appearance of political continuity (Fierke 2004; Edkins 2003; Kinnvall 2013, 154; Lerner 2020).

Each case study's war narratives thus must be read in the specific political frameworks dominant in the three communities at the time: in the Ukrainian case, in the spring of 2023, the second year of the war and a period of stalemate, attrition, and tentative hope for a powerful counteroffensive; in Syria, in the spring of 2019, and during the summer and autumn of 2020, when multiple factions were fighting for influence, causing high civilian death tolls and mass displacement particularly through the Idlib crisis; in the Palestinian sample, in the winter of 2020 throughout to April 2021, a period of mounting instability and the likelihood of the outbreak of large-scale violence in Gaza and parts of the West Bank.

Finally, while all war narratives have the same components, the retelling of the events and the portrayal of the victims and perpetrators by respondents differ. One apparent distinction is whether the events of war and its actors are framed in a 'political' or in an 'ethnic/sectarian' way. The differentiation between a 'political' or an 'ethnic/sectarian' rationale stems from the literature on nationalism and group identity,[1] however, I adjust the definitions so they can be applied to war narratives. While 'political' and 'ethnic/sectarian' are broad categories whose meanings differ in each case and specific context, the category of 'political war narratives' comprises a framing that focuses on differences in political ideas, power imbalances, and material interests as explanations for war. In contrast, what I call 'sectarian war narratives' centre their explanations around cultural ideas, ethnic (including language) and religious differences, and thus point to immaterial interests as the cause of war. Despite their context-specific nuances, a 'political' war narrative thus means that the respondent views the conflict and its parties as a result of political divisions, whereas a 'sectarian' war narrative is told from the perspective of someone who views the situation primarily as an ethnic/sectarian war. A typical political narrative assigns responsibility to different political players, parties, and interests. These dynamics are described in formal, political terms, often with hints to great power politics, authoritarian rule, a (corrupt) state and its security apparatus, or the military. Those who narrate the conflict in 'sectarian' terms tend to attribute the conflict to ethnic divisions (including race, heritage, language), and religious ideas.

[1] Borrowed from the literature on nationalism, a 'civic rationale' views group ties as based on shared ideas and political creeds, whereas an 'ethnic rationale' understands them as derived from primordial characteristics. Therefore, a 'civic rationale' centres around ideas of liberty, equality, or the concept of citizenship. In contrast, an 'ethnic rationale' focuses on religious beliefs, ethnicity, language, and cultural traditions (Eisenstadt and Giesen 1995, 80–82; Calhoun 2007; Reeskens and Hooghe 2010; Risse-Kappen 2010, 28 and 36; Valadez 2007).

Based on the theoretical building blocks of war narratives, this chapter presents Ukrainian, Syrian, and Palestinian accounts of war. In the questionnaire, respondents were allowed to freely formulate their narratives of traumatic events, assign responsibility to actors for the conflict, and identify its main victims. While each war story is unique, often detailed, and specific to the individual's experience and worldview, what follows presents aggregate data on the building blocks of war narratives alongside individual voices. Moreover, while taking the specific contexts of the sample groups into account, the narrative arcs are categorized as either 'political' or 'sectarian'. Illustrative quotes from the responses of Ukrainians, Syrians, and Palestinians allow those who have lived through these conflicts to speak for themselves.

Ukrainian Narratives of War

Ukrainians tell their narratives of war between March and May 2023. By that time, the war had entered its second year, and it had largely become a war of attrition, especially along the Eastern and Southern fronts. The survey period was also a time of 'stalemate'. One of the most intense battles during this time was in Bakhmut, a city in the Donetsk region. Russia, especially through Wagner Group forces, was trying to seize Bakhmut after months of fighting. Ukraine was defending it fiercely despite high casualties on both sides. Meanwhile, Russia regularly targeted Ukrainian energy infrastructure and cities with missiles and drones. At the same time, preparations for a substantial Ukrainian counteroffensive were underway: Western allies were supplying tanks, armoured vehicles, and increased training, but the offensive did not start until after the surveys were completed, in June 2023.

Thus, during the survey period, Ukrainian civilians were dealing with war trauma, housing and food insecurity, disrupted healthcare and education, as well as economic uncertainty. Furthermore, by May 2023, 6 million had fled Ukraine, with 3.7 million people internally displaced as of May 2023.[2] Internationally, Ukraine received increasing military, financial, and humanitarian support from the EU, US, and NATO, while countries like China and Brazil began to call more publicly for negotiations, though neither Ukraine nor Russia showed signs of compromising on core demands. The survey was therefore conducted at a moment of profound uncertainty combined

[2] UNHCR Annual Results Report 2023 Ukraine: https://reporting.unhcr.org/sites/default/files/2024-06/Europe%20-%20Ukraine%20ARR%202023_0.pdf (viewed 24 April 2025).

with cautious hope; although Ukrainians anticipated an impending majory counteroffensive, they were still subject to intense attacks and significant losses.

In this context, Ukrainian respondents' narrative of the ongoing war with Russia is predominantly one of Russian aggression and Ukraine's consequent defence of its national independence. In the sample, an overwhelming majority—92%—indicated that they perceive the outbreak of war as '*a consequence of Russia's aggressive foreign policy*'.

This broader sentiment is echoed in individual accounts. A twenty-eight-year-old man from Mariupol, now living as a refugee in Canada, states: '*The war started because Russia is an aggressor. This is dirty politics! Russia's and its aggressive foreign policy are to blame for it all*'. Similarly, a twenty-year-old woman from Crimea who fled to Germany tells her narrative: '*Russia wants to destroy us as a nation, subjugate us, and seize our lands. They have wanted to assimilate us and make us "proper Russians" for hundreds of years*'. A thirty-five-year-old woman from Kramatorsk who now lives in Lanivtsi in the Ternopil region reiterates this perspective, depicting the war as a result of

> Russia's desire to destroy the Ukrainian nation, combined with its confidence that the Ukrainian people would not mount strong resistance and that Ukraine would be quickly suppressed. Russia also believed that Western countries would not assist Ukraine, let alone provide long-term support, because they fear Russia and have their own economic interests. The war continues because of a lack of weapons in the Ukrainian armed forces, insufficient sanctions against Russia, and incomplete enforcement of existing sanctions. Therefore, Ukraine should be armed as much as possible, and pressure on Russia should be increased.

In contrast, only 3% of the Ukrainian respondents indicated that they perceive '*Ukraine's aspiration to join NATO*' as the main cause of war, and around 5% remain unsure (see Figure 3.1).

In her open response, a thirty-four-year-old woman from Kharkiv who now lives as a refugee in the EU, explained that she views the war as a consequence of Ukraine's aspiration to join NATO: '*The war started because of processes outside the country. This is not a war between Ukraine and Russia. This is a war between the West and Russia. And it started because we Ukrainians allowed ourselves to be used*'.

The war is a consequence of Russia's aggressive foreign policy	265	
The war is a consequence of Ukraine's aspiration to join NATO	8	
It's hard to say	14	

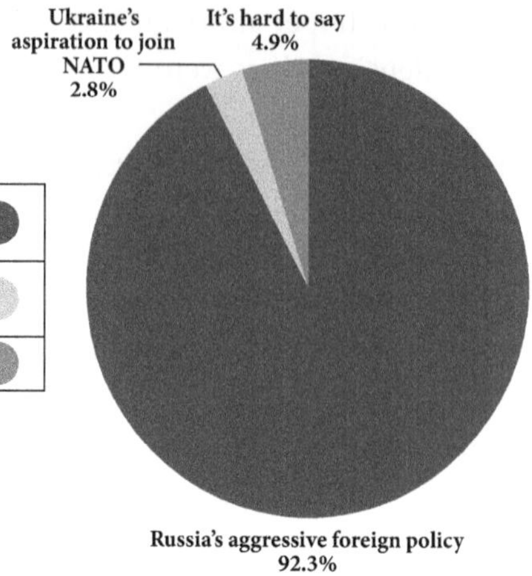

Figure 3.1 Perceived Causes of the War Among Respondents

Perpetrators and Their Victims

Any war narrative involves perpetrators and their victims. When Ukrainian respondents were asked to identify those responsible for the war, more than half (52%) assigned blame to both Russia's leadership and its citizenry. In contrast, 23% held only the Russian leadership responsible, while 7% blamed only the Russian citizenry. Altogether, 82% of Ukrainian respondents view Russia as the sole perpetrator of the war, while 18% identified other actors as also bearing some responsibility. Among these, 8% saw *both sides*—i.e., Ukraine and Russia—as responsible, and another 10% attributed blame to other international players (see Figure 3.2).

It follows that within the Ukrainian sample, 82% of respondents attributed responsibility first and foremost to Russia, whereas only 18% chose to attribute responsibility to '*others*'. Within the latter group, 7.7% assigned responsibility jointly to '*Russia and Ukraine*', 4% to '*Russia, Ukraine and International Players*', and 5% to '*Russia and International Players*', while merely two respondents (1%) attributed responsibility solely to '*Ukraine*'. As such, only a minority of 13% of respondents attributed at least some responsibility to the Ukrainian side. Exemplary of the few responses that took a critical stance towards Ukraine, a thirty-year-old man from Mariupol who fled to Wrocław wrote: '*The Ukrainian political elite, from the very beginning of independence, demonstrated weaknesses and a high level of corruption. At the same time, the Ukrainian national idea has been weak in some regions. Taken together, all this created favourable conditions for the invasion and the*

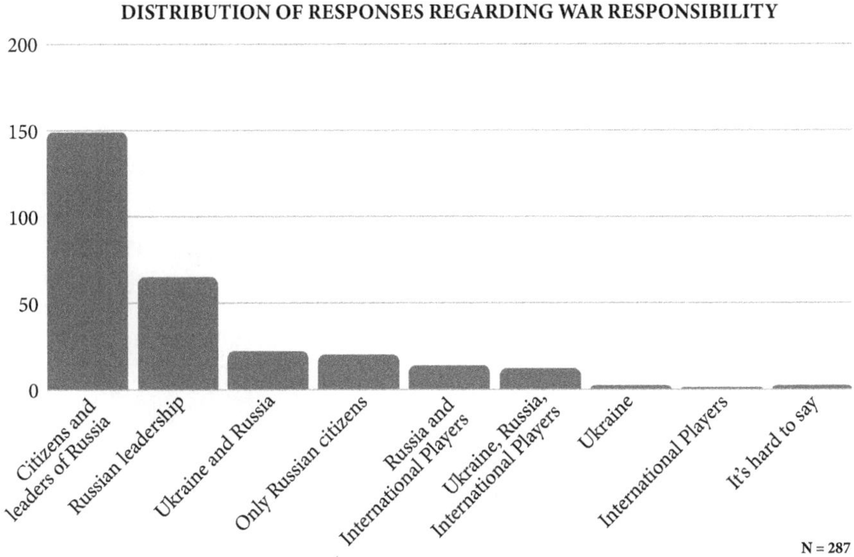

Figure 3.2 Perceived Responsibility for the War Among Ukrainian Respondents

start of the war. Having such an aggressive and dangerous neighbour as Russia, Ukraine could have avoided war only by developing a strong political elite that would have worked towards strengthening the national idea'.

Another crucial element of any war story is the question of who qualifies as a victim. In our sample of Ukrainians, a majority of 76% viewed '*all Ukrainians*' as the primary victim group of the ongoing war with Russia. For example, a twenty-three-year-old woman from Kharkiv explains that '*all Ukrainians suffer, because everyone has suffered material losses, psychological injuries, and some have experienced the pain of losing loved ones*'. Equally, a twenty-five-year-old woman from Kyiv, now living in Lviv, states that '*the entire Ukrainian people suffered. Both those who suffered greatly (lost property, relatives, and loved ones) and those whom the war did not affect so much. Especially children are suffering because they are the future generation, and they will be traumatized for the rest of their life*'. Along similar lines, a thirty-five-year-old woman from Kramatorsk, now living in Lanivtsi, explains: '*The entire Ukrainian people are victims of the war. People who were on the front line, who live under occupation, as well as military personnel who stood up to protect their native Motherland are especially suffering*'. Finally, a twenty-nine-year-old man from Nikopol, now living in Lviv, postulates: '*Clearly, all Ukrainians suffer. First and foremost, civilians whose homes and lives were destroyed or taken away; secondly, soldiers of the Armed Forces of Ukraine; and thirdly, civilians from*

Ukrainians	217	●
Direct victims	20	●
Children	15	●
Civilians	12	●
Ukrainians on territory	8	●
Soldiers	5	●
Ukrainians and Russia	3	●
Occupied	3	●
Western Ukrainians	1	●
Russian speakers	1	●
No victims	1	●
Humanity	1	●
Total	287	

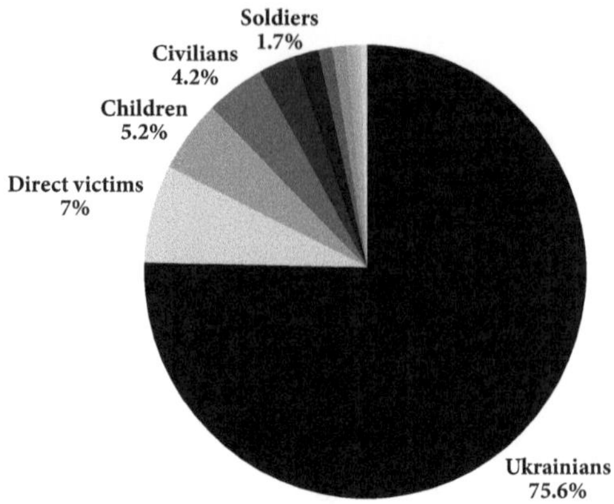

Figure 3.3 Designated Victim Groups According to Ukrainian Respondents

areas less affected by the war. Further down the line, we can also mention Europeans and Americans who sometimes must give up their usual way of life to help us. However: I don't consider Russians to be victims at all'.

While most respondents—despite varying degrees of emphasis—primarily identified Ukrainians as a collective victim, 24% exclusively singled out specific groups. These included those who had lost a loved one (*'direct victims'*, 7%), children (5%), civilians (4%), Ukrainians who remained in the country during the war (3%), soldiers (1.7%), and those living under occupation (1%) (see Figure 3.3). For instance, a twenty-four-year-old respondent from Vasylkivka village in the Dnipro region stated that the primary victims of the war are those '*who suffered and lost their families and homes. All those soldiers who died defending our land*'. Yet even these answers embed individual victims within the broader Ukrainian collective. This also applies to respondents who self-identify as Crimean Tatars. As a twenty-seven-year-old woman from Simferopol explained:

> The citizens of Ukraine are undoubtedly the main victims of this war; especially those who lost their lives and loved ones. This also includes those who cannot get to Crimea, to their own homes and who have been persecuted there since 2014. Ultimately, all those who consider themselves Ukrainian have become victims—some lost their lives, some lost their homes, and some lost their freedom.

Taken altogether, the survey results suggest that the typical Ukrainian narrative of war consists of two elements: the view of Russia as the perpetrator, and the view of the Ukrainian collective as the victim. The response of a twenty-five-year-old man from the Donetsk region, now living as a refugee in Estonia, puts this narrative in stark terms: '*The war started because of Putin's imperialist and revanchist sentiments already in 2014. Putin seeks the restoration of the Soviet Union. The second phase of the war began in 2022 because Putin thought he could take Ukraine within a matter of days*'. In his opinion, '*the war is the result of Russia's aggressive foreign policy, and responsibility rests* first and foremost *with the leadership of Russia. The main victims of this war are Ukrainians*'.

The Ukrainian Narrative Is Political

The average response also illustrates that, in Ukraine, the war is predominantly viewed through a national and political lens. When respondents were asked to indicate the goals of Ukraine's war effort, the majority cited terms such as *independence*, *freedom*, *sovereignty*, and *territorial integrity* (see Figure 3.4). In other words, the prevailing view is that Ukrainians are defending their nation against Russian aggression, as reflected in the 88.9% of responses coded as 'independence'.

While these responses highlight a strong attachment to Ukrainian territory and national borders, for many respondents, *independence* and *sovereignty* also signify a closer alignment with the West—that is, with Europe, democracy, and the rule of law:

> Ukrainians are fighting for the right to exist as a nation—for the pro-European choice that was won through the blood of people during the Revolution of Dignity on Maidan Nezalezhnosti. We are also fighting for the universal principle of territorial integrity, which has been violated by Russia. We fight for the right to join the EU and NATO—or not. The point is the freedom to choose for ourselves. For the right, as Shevchenko wrote, to have our 'cherry orchard by the house' and live in peace. For the right to be a democratic state with a responsible political elite (Response from an eighteen-year-old man from Kyiv, now living as a refugee in Blain, France).

Similarly, a twenty-one-year-old woman from the Rivne region, now living in Kyiv, explains: '*Ukrainians are fighting for the sovereignty and territorial integrity of our state, for its freedom and independence—for a future free from occupiers*'. To fight for national independence, in the words of a twenty-nine-year-old woman from Kyiv who now lives as a refugee in Germany, means to fight '*for the right to be Ukrainian and not "Russian", because Russia erases the national identity of other peoples. For values such as democracy, freedom of speech, freedom of self-expression, and self-governance*'.

Independence	255	N = 287
Europe	19	
Free world	8	
Anti-War	5	

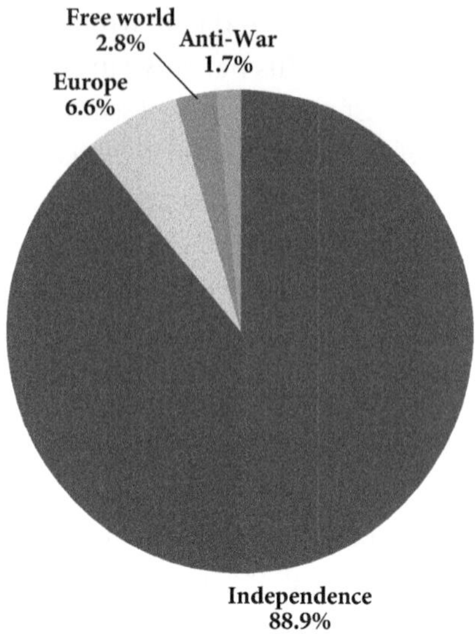

Figure 3.4 Perceived Goals of Ukraine's War Effort

While this point is likely implied in most answers, only 6.6% of respondents explicitly mentioned that Ukrainians are fighting not just for *any* national future, but for a European one. As a thirty-seven-year-old woman from Odesa, now living as a refugee in Kirkcaldy, Scotland, explains: '*Ukrainians are fighting for the right to build their state based on European values—on the principles of democracy, respect for citizens, human dignity, equality, and freedom*'.

In conjunction, the different components of the Ukrainian war narratives indicate that respondents predominantly view the events through a 'political' lens. In the opinion of our sample group, Russia's aggressive foreign policy, its authoritarian ruler, the '*brainwashed*' Russian citizens, and major power imbalances bear responsibility for the war. Notably, in the Ukrainian context, an 'ethnic/sectarian' logic is rooted in ideology rather than ethnicity, dividing Ukrainians along an East–West fault line. Furthermore—also largely because of the war—independence has become interwoven with closer association with '*the West*', reinforcing a Ukrainian identity that is distinct and separate from Russia. In this context, a neat separation of war stories into political or sectarian narratives does not fully apply. However, some responses do include more ethnic or sectarian language than others. For instance, a thirty-two-year-old man from Poltava, now living in Chernihiv, offered the following response:

> The Muscovites' [a derogatory term used often by respondents for Russians] have decided to calm their phantom pains caused by the collapse of the so-called superstate of the USSR and to restore this state entity—a neo-empire—through forceful methods. Russians largely support the idea of their leader because they consider Ukrainians inferior and unworthy of having their own state. This is called 'Rashism'.

Overall, the war narrative of the Ukrainian respondents is relatively uniform and follows an antagonistic, political logic that positions Russia as the aggressor and Ukrainians as defenders of Ukraine's independence and national integrity in response to external aggression. Responsibility for the war is overwhelmingly attributed to Russia; however, quite notably, this responsibility is not assigned solely to Russia's political leadership, but also to the Russian citizenry. It follows that, because of the war, all Ukrainians have suffered. Across most respondents' answers, a heightened sense of collective victimhood prevails—one that unites Ukrainians in standing strong against an overpowering enemy.

Syrian Narratives of War

The Syrian respondents tell their narratives of war between May and June 2019, and between July to October 2020. By that time, the Syrian civil war was well into its ninth and tenth years. The situation was highly fragmented with control divided among multiple factions, including the Syrian government led by Assad and backed by Russia and Iran, with opposition groups including Turkish-backed factions, the Kurdish-led Syrian Democratic Forces (SDF) and Jihadist groups such as Hay'at Tahrir al-Sham (HTS), as well as a number of foreign actors, primarily Russia, Turkey, the US, and Iran.

The first survey period, May to June 2019, coincided with the so-called Idlib crisis in which the Assad regime, supported by Russian airstrikes, launched an offensive to retake rebel-held Idlib Province, prompting high civilian death tolls and waves of displacement. By comparison, the second period of the survey, July to October 2020, was relatively quieter, following a ceasefire in Idlib in March 2020, brokered by Turkey and Russia. However the conflict was far from resolved, the front lines remained active, and clashes continued. Moreover, living conditions had worsened for Syrians in 2020: a looming economic crisis had rendered basic goods such as bread unaffordable, while the COVID-19 pandemic further strained an already obsolete healthcare system.

The Syrian respondents are thus recounting their war narratives while experiencing new waves of displacement amid violence in Idlib and north-eastern Syria, a shifting political and military landscape, a collapsing economy and healthcare system, exacerbated by the COVID-19 pandemic, and profound uncertainty about the future and the possibility of return for the displaced.

In this context, most respondents describe the Syrian war in terms of dictatorship, political interests, and external interference. For illustration, take this answer from a thirty-seven-year-old man from Hama who now lives as a refugee in the UK:

> The conflict started when some Syrians found the courage to rebel against dictatorship, hoping for a better future. They took to the streets in peaceful demonstrations, which were quickly met with force. Peaceful citizens aspiring for freedom and democracy were killed by the regime or arrested and tortured. Islamists found an excellent opportunity to implement their project. They were used by the Syrian regime and its allies—Iran and Russia, either directly or indirectly—to undermine their struggle against dictatorship. They were also used by foreign countries, namely Turkey, Qatar, and Saudi Arabia, as part of their power competitions and proxy wars. The Islamization of the political struggle was then exploited by the regime to scare parts of Syrian society away from any potential change. This led to the widening of cleavages within Syrian society—a project the regime had been working on for decades. The regime sat back, watching as different parts of society turned against each other, while those who demanded freedom became irrelevant and were forced to leave the scene. As people became cannon fodder, and as foreign powers—namely the US and Russia—showed little will to force an end to the conflict, Syria became just another victim of a dysfunctional world order. Thus, the conflict goes on and on.

Perpetrators and Their Victims

When the Syrian narratives of the war are broken down into the building blocks of war stories, a majority of respondents in the Syrian sample (49%) assigned responsibility for the ongoing conflict to both the Assad regime and the interference of international actors. Meanwhile, 37% attributed responsibility solely to the Assad regime. For instance, a thirty-eight-year-old man from Homs, now living as an internally displaced person (IDP) in Afrin, explained: '*The cause of the conflict is a criminal regime based on systematic killing, displacement, and widespread administrative corruption throughout the country's institutions. This regime has deprived the people of the most basic rights and has monopolized the nation's wealth, concentrating it in the hands of those close to the ruling authority'.* Along similar lines, a thirty-nine-year-old man from Haseka, now living as a refugee in Germany, states: '*The cause of the conflict is corruption, tyranny, exclusion, and the absence of even the most basic standards of human rights. The reason it continues—alongside the regime's ideology—is the emergence of extremist Islamic currents whose fanaticism is no different from that of the regime itself*'.

While an overwhelming majority of responses point to Assad's regime as bearing the brunt of the responsibility for the Syrian war, only one respondent assigned responsibility to the opposition, while four respondents (2.5%) assigned responsibility to both parties in the civil war: the regime and its opposition (see Figure 3.5). To give an example, a sixty-four-year-old man from Latakia asserts: '*Both the opposition and the regime are equally responsible—along with the countries that support them*'.

When stories of war assign responsibility, they also designate a victim group that suffers as a result of the perpetrator's actions. In the Syrian sample, 62% named '*all Syrians*' as the primary victims of the Syrian war. The remaining 38% identified specific groups as suffering the most: Sunnis (12.5%), minorities (12%), the opposition (9%), and children (4%) (see Figure 3.6).

Here are excerpts illustrating the majority opinion that designates all Syrians—that is, ordinary civilians—as the main victims of the war. A sixty-four-year-old woman from Damascus, now living as a refugee in the US, states: '*The Syrian people suffer. They face the destruction of their homes and displacement, the killing of their sons, and the rape of women*'. Similarly, a sixty-five-year-old woman from Qamishli, now living as a refugee in Australia, explains: '*The average Syrian citizen suffers because this conflict has robbed them of their life*'.

International players	17	●
Opposition	1	●
Opposition and Regime	4	●
Regime	59	●
Regime and international players	78	●
Total	**N = 159**	

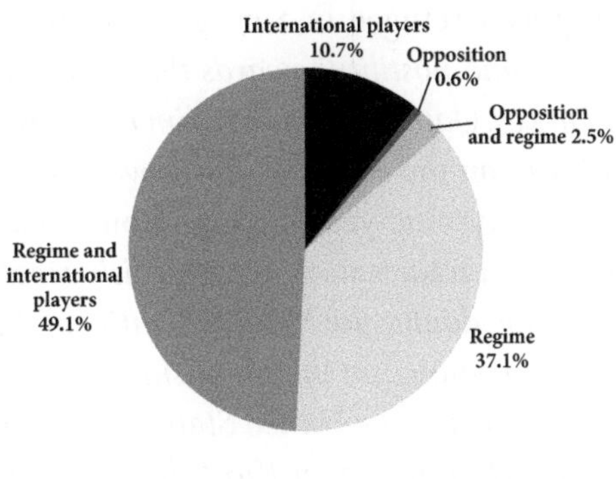

Figure 3.5 Perceived Responsibility for the War Among Syrian Respondents (N=159, contains only 2019 sample)

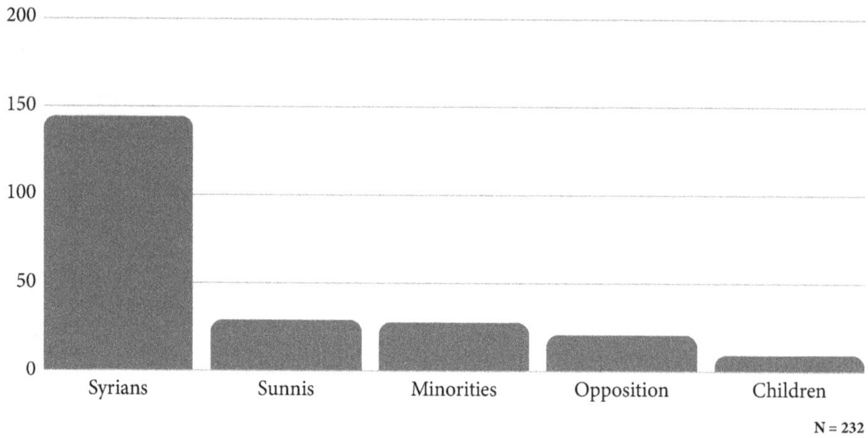

Figure 3.6 Designated Victim Groups According to Syrian Respondents

Or take this response from a thirty-seven-year-old man from Homs, now living as a refugee in London: '*Ordinary people are the ones who are suffering. First, those who have been displaced internally or forced to flee to neighbouring countries; then, those who are still inside Syria. Within society, children are suffering the most, followed by women and young men, who face the risk of being conscripted or drawn into violence*'. Similarly, a sixty-one-year-old man from Rif Dimashq, now living as a refugee in Turkey, states: '*Most Syrians suffer. Some suffer death, others from injury, poverty, displacement, or asylum. But the detainees in the regime's prisons, wishing for death to end their suffering, are the most affected*'.

Among the respondents, 12% explicitly mentioned minority groups as the main victims—specifically the Kurds, but also Alawites, Ismailis, Druze, Christians, and Yazidis. As a sixty-one-year-old man from Hasakah, now living as a refugee in Sweden, explains: '*The Kurds are the main victims. Racism and hostility towards the Kurds have united the opposition supported by Turkey, Qatar, and the Muslim Brotherhood on the one hand, and the regime and Iran on the other hand—as well as parts of Syrian society*'. Along similar lines, a fifty-nine-year-old man from Rojava, al-Hasakah, now living in South Kurdistan, Iraq, states: '*The Kurdish people are suffering the most because the opposition doubts the Muslim identity of the Kurds, and Turkey exploits the situation to pounce on them. Russia encourages and endorses of Turkey's actions in order to keep the United States at bay. But Turkey has no other goal in Syria than the elimination the Kurdish people*'. That Turkish involvement in the conflict has rendered the Kurds of Syria the main victims is also asserted by a sixty-one-year-old man from Al-Hasakah, who lives as an IDP in Qamishli:

'The Kurds suffer the most because they are under Turkish occupation. They are forcibly displaced from their areas, subjected to killings and property seizure, and face demographic change in their regions'.

While Syrian Kurds tended to single out their own ethnic group as the main victims of the war, some respondents followed a sectarian logic and identified Sunnis as the primary victims. For instance, a fifty-eight-year-old man from Rif Dimashq, now living as a refugee in Germany, states:

> Sunni Arabs are the ones who suffer the most because the regime sees them as political opponents. As a result, they kill, destroy, and steal from them. Syria today is not a place where one can live a dignified life. It is a hotbed for pro-regime groups, who can do whatever they like. When they are hungry, they devour everyone's dignity. No one is safe from the regime's cruelty. And I cannot even begin to descri-bethe suffering of detainees in prisons, deprived of any rights, and the pain of their families.

Along similar lines, a sixty-five-year-old man from Rif Dimashq, who fled to Lebanon as a refugee, explains: *'The victims are all the Syrian people. But those who have paid the highest price are the ones living in regions that were robbed, bombed, destroyed, and displaced. Most of these areas were inhabited by Sunnis. However, even the Alawites were not spared, as they were forced to fight and kill in revolutionary-held areas to detain and suppress protesters'.*

Others, however, followed a political logic and singled out the opposition as the main victims. For instance, a fifty-year-old man from Daraa, Busra al-Sham, now living as a refugee in Germany, exclaims: *'The main victims are the revolutionary people who stood up against the regime'.* A forty-two-year-old man from Aleppo, now living as a refugee in Denmark, suggests that the main victims are *'those with liberal political views who believe in the values of democracy and coexistence. These people are targeted by all parties participating in the civil war'.*

Overall, in the context of the Syrian civil war—and as shown in Figure 3.6—while the majority of respondents identified all Syrians as the main victim group, many also singled out specific ethnic and sectarian communities as primary victims of the conflict.

Political and Sectarian Narratives of the War in Syria

Following the dissection of the Syrian narratives into assigned responsibility and designated victims, the accounts were analysed according to whether they followed a political or ethnic/sectarian logic. In a political narrative of the Syrian war, the respondent describes the conflict and its actors in political terms. In contrast, a sectarian narrative frames the conflict primarily as an ethnic or sectarian war.

A typical answer that followed a political rationale assigned responsibility to various political players, parties, and interests. These were described in formal political terms, often with reference to authoritarian rule, the state and its security apparatus, or the military. Moreover, many responses described the war in terms of its internationalization. Such narratives were coded as 'political' when they invoked the language of great power intervention, regional competition for influence, or political interests involving the intervention (or non-intervention) of the international community. Furthermore, those who narrated the conflict in political terms often stated—when asked about the primary victims—that *'all Syrians'* or *'ordinary citizens'* were suffering the most as a result of war and forced displacement.

In contrast, when the war was presented in ethnic terms, a specific ethnic group—usually Muslim Sunnis—was singled out as the primary victim group. Respondents who described the war as the result of an ethnic or sectarian agenda typically assigned responsibility to an *'Alawi-minority tyranny'* over a *'Sunni majority'*. If they considered the international dimension of the conflict, they characterized international players as Shiite, Sunni, Islamic, Western, or Judeo-Christian actors, each pursuing its own ethnic or sectarian agenda.

As illustrated in Figure 3.7, a clear majority of respondents (70.7%) followed a political rationale in recounting what had happened in Syria since 2011. For example, a fifty-year-old man in Aleppo explains: *'The Syrian regime and the Ba'ath Party refused to respond to the* - a political narrative: *'Everyone is after their own personal and financial gains, and none of them care*

Political	118	●
Sectarian	49	●
Total	**N = 167**	

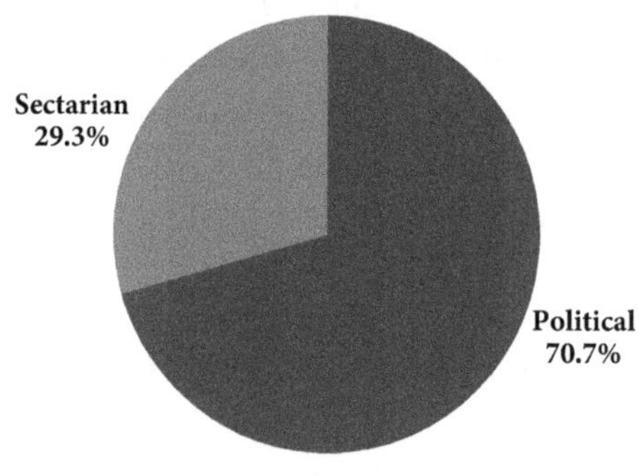

Figure 3.7 Political and Sectarian Narratives of the Syrian Conflict

about the interests of the people. Therefore, the people—the Syrian people, the children, the middle class, and the poor—are the main victims of this conflict'.

Like these two, most responses that framed the conflict in political terms pointed to the state's security apparatus, its army and intelligence services, and the intervention or non-intervention of other countries and the international community as drivers of the Syrian conflict. In the words of a fifty-year-old man from Damascus: '*Firstly, the regime is responsible for the escalation of the conflict because it had the power, the influence, the money, and the intelligence in 2011. Yet its policies led to violence and foreign intervention*'. A forty-two-year-old man from Damascus, who follows a similar logic, adds: '*The ordinary people who did not align themselves with any of the conflict's parties are suffering the most! The regime, Russia, Iran, the USA, and the Gulf countries—all act in their own interests. Bribery and nepotism are rampant. Loyalty should not be to the president—it should be with the land and the people!!*'

As such, when assigning responsibility for the conflict, most respondents who narrated it in 'political' terms blamed the ruling regime for escalating the violence and foreign intervention for prolonging it. Among the most frequently named foreign actors were Iran (and affiliated Shiite militias such as Hezbollah) and Russia, followed by the Gulf countries and, to a significantly lesser extent, the United States. While opinions varied about these players, a clear majority of respondents equated foreign involvement with the spread of sectarian strife. As a result, sectarianism was seen not as something deeply rooted within Syria, but as an external threat imported from abroad. Their story of the Syrian conflict was not one of primordial ethnic or sectarian hatred, but rather one grounded in political interests that intensified both internal divisions and international entanglements.

In contrast to the majority of Syrian respondents who followed a political rationale in their narratives of the Syrian war, 29.3% pursued an ethnic or sectarian rationale. To give an example of such a narrative, a sixty-seven-year-old man from Raqqa shares his views on the war:

> The Sunni majority has paid the highest cost in this war in terms of counting killings, displacements, and imprisonments. Arab Sunnis are therefore the group that has suffered—and continues to suffer—the most, both in the past and today. The responsibility lies with the Syrian regime, which has held on to power since 1970, embodied by the Assad family, its leadership, the Ba'ath Party, the Alawite sect, and the vast majority of businessmen who have supported the regime.

A sectarian version of the Syrian civil war thus depicts the conflict as driven by religious and ethnic dominance, with sectarian identity seen as a central force. As a forty-two-year-old man from Aleppo, now living as a refugee in Denmark, explains: '*In summary, the conflict was caused by a political crisis stemming from the domination of a certain group over the country's resources. It then the conflict turned into a sectarian and national war between Sunnis, Alawites, Kurds, Arab nationalists, and Islamic extremists*'. Along similar lines, a fifty-five-year-old man from Jasim in Daraa Governorate, now a refugee in Jordan, states: '*In short, the Syrian war is a struggle against tyranny, injustice, and persecution carried out by one sect against Syrian society as a whole. The reason for its continuation is the injustice of the international community, its lack of humanity, and its failure to act in the face of the Assad gang's massacres of the Syrian people*'. Finally, a sixty-five-year-old man from Qamishli, now living as a refugee in Germany, describes the conflict in stark ethnic and sectarian terms: '*The actual reason for the war is the racist Arab mindset and a tyrannical Islamic opposition*'.

Overall, the war narrative retold by Syrian respondents centres on a ruthless dictatorship, competing political interests, and the Syrian people's struggle for freedom. However, interpretations of these elements vary slightly among respondents. The majority framed their responses through a political lens, identifying abuses of power and political interests as the primary causes of the conflict, while 30% interpreted the war through an ethnic or sectarian lens. Almost all respondents in this sample held the Assad regime responsible for initiating the war, and most also attributed blame to various international actors for escalating it. While the majority described these actors' roles in political terms, a significant portion (30%) emphasized sectarian motivations as the primary drivers. Although the idea of collective victimhood—encompassing all Syrians—emerged as the dominant narrative, whenever a specific group was identified as the main victim, it was most often an ethnic or sectarian group. Politically neutral or broadly defined groups, such as women and children, were less frequently singled out in this way.

Palestinian Narratives of Conflict

Palestinian respondents completed the survey between November 2020 and April 2021. In the West Bank and Gaza, this period was marked by mounting instability and a growing likelihood of large-scale violence, although major hostilities had not yet resumed. The previous year had witnessed significant political developments, most notably the Trump administration's 'Peace to Prosperity' plan, which deepened existing asymmetries by legitimizing Israeli sovereignty over substantial parts of the West Bank. The plan was overwhelmingly rejected by Palestinian actors, reinforcing perceptions of a de facto annexation. Meanwhile, the COVID-19 pandemic intensified the humanitarian crisis, compounding restrictions on movement and economic activity across the Palestinian Territories.[3]

Nevertheless, the lived realities of respondents in Gaza and the West Bank diverged markedly in the survey period. In Gaza, residents endured a protracted humanitarian crisis characterized by blockade, recurrent low-intensity conflict, systemic economic deprivation, and chronic infrastructure failures. In the West Bank, Palestinians faced intensified political repression, settler violence, settlement expansion, and increased confrontations with Israeli security forces, especially in East Jerusalem.[4] Shortly after the survey concluded, Israel launched *Operation Guardian of the Walls* in May 2021. The survey therefore captures Palestinian experiences at a critical juncture, when structural injustice and political conflict were converging towards open violence. This occurred amid widespread disillusionment with the prospects for negotiation and a growing sense of abandonment by the international community following shifts in US policy.

In this context, most Palestinian respondents describe the Israeli–Palestinian conflict as a Zionist occupation of Palestinian lands supported by Western countries. The following response from a forty-year-old man from Gaza, who identifies as a Palestinian refugee originally from Sarafand al-Kharab village, is an illustrative example:

> The cause of the conflict is the occupation of Palestinian land and the plunder of its assets by the Israelis. Israelis deny Palestinians access to their homeland and its goods. The main responsibility thus lies with the Israeli occupation. However, historical responsibility is also shared by Britain, as Britain granted permission for the establishment of the Israeli state. Finally, the United States also bears responsibility for the conflict as it supports Israel militarily, economically, and politically.

[3] United Nations Office for the Coordination of Humanitarian Affairs (OCHA). 2021. *Occupied Palestinian Territory: Humanitarian Response Plan 2021*. Available at: https://www.un.org/unispal/wp-content/uploads/2020/12/OCHAHUMRESPLAN_151220.pdf (viewed 26 April 2025).

United Nations Relief and Works Agency for Palestine Refugees in the Near East (UNRWA). 2020. *Annual Operational Report 2020*. Available at: https://www.un.org/unispal/wp-content/uploads/2021/08/UNRWAANNUALTPT_100821.pdf (viewed 26 April 2025).

[4] United Nations Relief and Works Agency for Palestine Refugees in the Near East (UNRWA). 2021. *Emergency Appeal 2021*. Available at: https://www.un.org/unispal/wp-content/uploads/2021/03/UNRWAEMERAPPEAL_020321.pdf (viewed 26 April 2025).

Perpetrators and Their Victims

In their narratives of the conflict, many Palestinian respondents attribute responsibility to Israel. In total, 43% of respondents said that Israel is to blame—along with, by extension, the Western powers that support it. Most frequently mentioned are Britain (in the past) and the United States (in the present). For instance, a thirty-seven-year-old man from Gaza, who identifies as a Palestinian refugee from Burqa, wrote:

> The conflict began when Britain granted the Zionist Jews the lands of Palestine, donating what it did not own to those who were not worthy. The conflict continues because of the ongoing plunder of these lands, and Britain bears primary responsibility for this injustice inflicted on the Palestinians more than 70 years ago. The so-called Israel is the primary perpetrator, and those who support and protect its presence—such as the United States, Britain, and Arab countries like Egypt and Jordan, which help to secure Israel's borders—also bear responsibility for this conflict.

However, an equal number of respondents—also 43%—said that both sides, Israelis and Palestinians, share responsibility for the conflict. These responses often included criticism of internal Palestinian divisions, particularly between Hamas and Fatah. For example, a twenty-one-year-old man from Ramallah, who identifies as a Palestinian refugee from Yaffa, wrote:

> The cause of the outbreak of conflict was the British mandate that promised the Jews a state in a land that was not theirs, while ignoring the rights of the indigenous people. Although it was possible to receive the Jews as immigrants and give them Palestinian nationality and full civil rights, the conflict has persisted due to Jewish greed for land and the refusal to accept a peace plan in which the Palestinian people would be recognized as a people who lived—and still live—on the land, and who have the right to return to the land from which they were exiled. Responsibility therefore lies with America and Israel, but also with the Palestinian Authority. The Palestinian Authority is too weak diplomatically; furthermore, Hamas and Fatah are unable to resolve their division. Arab politicians are also to blame—they are weak and inconsistent in their support for the Palestinian cause.

Similarly, a forty-one-year-old man who identifies as a Palestinian refugee from Bayt Tima village and now lives in Gaza wrote:

> The blame lies with the Zionist Western colonial project, which has usurped Palestinian land. It is responsible for the displacement of the Palestinian people, both

within and outside Palestine. The reason for the continuation of the conflict is the American–European bias in favour of Israel at the expense of Palestinian rights. The perpetrators are Britain, through the Balfour Declaration; U.S. imperialism; the Israeli occupation; and finally, the parties to the internal Palestinian division—Fatah and Hamas.

Finally, a thirty-five-year-old woman from Yaffa, who identifies as a Palestinian refugee living in New York, believes that the conflict has persisted due to '*overall stubbornness on both sides. Neither side is willing to accept any concessions, and the international community is overwhelmingly biased. Both sides are at fault, and pointing fingers at any one side won't change anything*'.

It is notable that while 43% of respondents in the Palestinian sample attribute responsibility either to both sides or to the opposing side (Israel), 13% place the blame solely on the Palestinian side for the absence of peace (see Figure 3.8). These respondents point to the weakness of the Palestinian Authority—particularly the division between Hamas and Fatah, and between Gaza and the West Bank. They also blame fellow Arab and Islamic countries, especially Jordan and Egypt, which are described as '*traitors*' or '*corrupt dictatorships*' due to their peace treaties with Israel.

While the Palestinian responses are almost evenly split in their assignment of responsibility to Israel alone or to both the Palestinian Arab and Israeli sides, they are fairly uniform in their designation of the main victims.

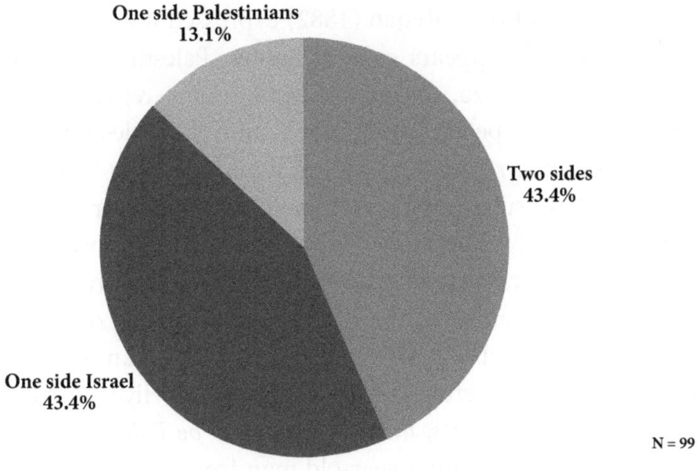

Figure 3.8 Perceived Responsibility for the War Among Palestinian Respondents

Palestinians	81	
Refugees	10	
Gazans	7	
Martyrs	1	
Total	**99**	

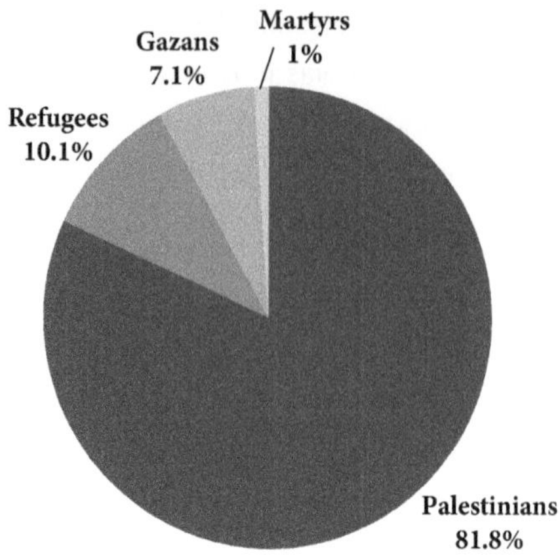

Figure 3.9 Designated Victim Groups According to Palestinian Respondents

A total of 82% identified all Palestinians as the primary victims of the ongoing conflict (see Figure 3.9). As a forty-year-old man from Gaza, a Palestinian refugee originally from Lod, stated: '*Everyone who considers oneself Palestinian suffers*.' A twenty-two-year-old woman from Ramallah wrote: '*There is no Palestinian who does not suffer, regardless of whether he is a refugee or in his country*.' Similarly, a thirty-seven-year-old woman from Nablus, now living in Ramallah, identified all Palestinians—'*the Palestinian people, whether on the land or in the diaspora*'—as the main victims of the conflict.

With 82% of responses naming all Palestinians as victims, a strong sense of 'shared suffering'—as Ernest Renan (1882) expressed it in his famous lecture *What Is a Nation?*—appears evident among Palestinian respondents across the West Bank, Gaza, and the diaspora. The remaining 18% of respondents singled out specific subgroups within the Palestinian population, such as '*Palestinian refugees*' (10%), '*Gazans*' (7%), or '*martyrs*' (1%) (see Figure 3.9). These subgroups are defined by political or geographical status and were typically named by respondents who themselves identified with them. However, even when specific subgroups were singled out, respondents ultimately embedded their suffering within that of the national collective. This is illustrated by a thirty-five-year-old woman who identifies as a third-generation 1948 refugee from Yaffa and now lives in New York, who wrote: '*Palestinian refugees, those in exile... to be Palestinian is to suffer, wherever you are*.' A forty-nine-year-old man from Gaza, who identified himself as a 1948 Palestinian refugee from the village of Al-Jiyeh, similarly stated:

> Refugees in diaspora countries—especially those in refugee camps in Lebanon, Syria, and elsewhere—suffer the most. Next are refugees in the camps of the Gaza Strip and the West Bank, followed by Palestinians who live in the cities of the West Bank and the Gaza Strip, and then those who live in the occupied interior since 1948. Ultimately, every Palestinian, wherever they are, suffers from the consequences of the Israeli occupation.

Interestingly, some respondents stressed that Palestinians suffer not only under Israeli occupation but also under the mismanagement of the Palestinian Authorities. These respondents perceive Palestinians as victims of their own elites. A twenty-five-year-old man from Birzeit, West Bank, claimed: '*Most of society suffers—except for the corrupt, politicians, the bourgeois, and those affiliated with or working with the Authority*. Similarly, an eighteen-year-old man from Al-Arroub refugee camp in the West Bank asserted: '*The people as a whole suffer, but the leadership does not; they live their best lives while poor people have no means of living*'.

What stood out within the Palestinian sample was that no participant explicitly identified Israelis as victims of the conflict. Only one nineteen-year-old woman from Gaza, now living in the Palestinian diaspora in the Netherlands, articulated a universalist understanding of suffering that implicitly encompassed all parties to the conflict. However, this framing did not extend to recognizing Israelis as victims. Drawing on a liberal perspective, she stated: '*All humans—everyone involved—suffer, even the oppressors. But most definitely, Palestinians suffer the most . . . . Just go to any Palestinian home. You will witness the trauma, the despair, the sadness, the hopeless hopefulness.*'

Political and Sectarian Narratives of the Conflict between Israel and the Palestinians

While any war narrative designates perpetrators and their victims, the characterizations and rationale behind the narrative arc can follow either a 'political' or an 'ethnic/sectarian' logic. A typical political logic presents the Israeli–Palestinian conflict as a territorial, national, or political dispute—as a clash of political interests over the same land, a result of power politics and asymmetries, or a failure of political leadership (on both sides). In contrast, a sectarian logic describes the conflict as rooted in religious and ethnic divisions, governed by an *us versus them* worldview, and driven by hatred, racism, and extremist ideologies.

NARRATIVE OF WAR

Political	83	●
Sectarian	21	●
Total	**N = 104**	

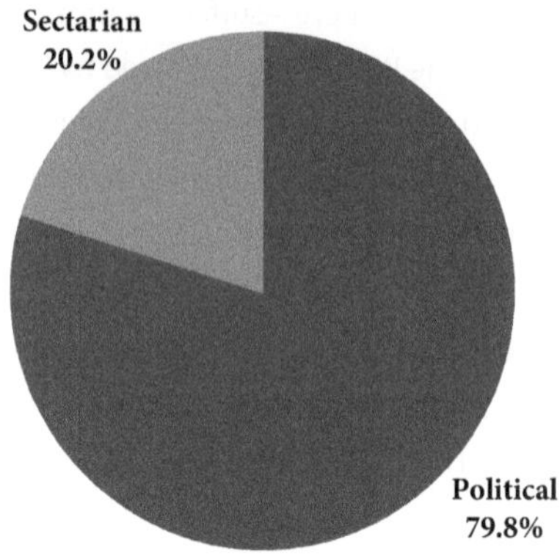

Figure 3.10 Political and Sectarian Narratives of the Israeli–Palestinian Conflict

Within the Palestinian sample, 80% framed the conflict with Israel as a political one, whereas 20% described it as sectarian, religious, or ethnic in nature (see Figure 3.10). Take, for instance, the response of a twenty-two-year-old woman from Hebron, who describes the conflict as political: '*The conflict inside Palestine is a scattered struggle between leaders, governments, and different movements. The conflict between Israelis and Palestinians revolves around leaders, money, and interests. As for its historical roots, the main reason lies in the history of Palestine, whether religious or geographical in nature. Unfortunately, Palestine itself is not united, and so the occupation continues, while the Palestinian people continue to suffer*'.

Similarly, a twenty-two-year-old woman from Bethlehem, who identifies as a Palestinian refugee from Zakariyya village, frames her response in political terms:

> The history of the conflict is rooted in the Zionist movement, which interpreted the Balfour Declaration as 'a land without a people for a people without a land'. The reason for the conflict also lies in the religious importance of Palestine and the significance of its geographical location. Those in power are responsible for the ongoing conflict, as they misuse their power and support the wrong causes. The victims are the youth of Palestine—the current generation deprived of the ability to dream and live like young people in the rest of the world.

In contrast, a twenty-six-year-old woman from Gaza, who identifies as a Palestinian refugee from Al-Majdal, explains the conflict in sectarian terms:

> The conflict is religious and existential: the Zionists fight all those who are not of their religion, and they kill Palestinians every day, in cold blood, to steal the land

> and take it from us. They want to destroy the Al-Aqsa Mosque and rebuild their Temple, but they have no place in our homeland. We are the owners of the land, and they are mere outcasts who were expelled from the West. Yet the entire Arab world has failed Palestine and has not supported us. Instead, Arab rulers have normalized relations with the Zionists. America, as the staunchest supporter of Zionist fascism, is first and foremost responsible for our current situation. It supports them financially and militarily and is thus responsible for the theft of Palestinian land. Those who suffer most are the Palestinian refugees, scattered across the world—in Gaza, Lebanon, Jordan, Syria, and beyond.

Overall, the war narrative as told by our Palestinian sample follows a story arc that begins with Israeli occupation and culminates in Palestinian loss and collective suffering. While 43% assigned full responsibility for the conflict to the Israelis, a majority of respondents (56%) also acknowledged the Palestinian role in the ongoing conflict, though always with reference to the clear power asymmetries between the two sides. A strong sense of collective Palestinian victimhood is evident across responses from the diaspora, the West Bank, and Gaza. Wherever subgroups were singled out as specific victims, they were characterized by political disadvantage within the broader Palestinian collective. Importantly, however, most Palestinian respondents (80%) framed the conflict as a modern political and territorial struggle, rather than as one rooted in ancient sectarian divisions.

Conclusions

Mapping the narratives of war of Ukrainian, Syrian, and Palestinian respondents, the analysis reveals the following trends.

In the Ukrainian case, the account of the war with Russia is rather uniform and in a majority of 92% follows an antagonized, political logic that positions Russia as the aggressor and Ukrainians as defenders of the country's independence and national integrity in reaction to this external aggression. As a consequence, a majority of 82% distributes responsibility solely to Russia, while a minority of 18% includes responsibility on the part of Ukraine and other international players. Notably, when assigning responsibility to Russia, most respondents make clear that by Russia they mean the leadership and its citizens. Finally, in Ukraine, a heightened sense of collective victimhood is evident, with 76% naming the Ukrainian collective as the primary victim of Russia's aggression. The remaining 24% single out a specific group that is usually politically defined, that is, those living in the occupied territories or those who are fighting on the front lines.

In the Syrian case, respondents' account of the war is a little less uniform. Nevertheless, a majority of 71% follows a political logic when recounting the war. These respondents describe the civil war as a result of dictatorship paired with political interests and the fight of the Syrian people for freedom. In contrast, 29% describe it in ethnic-sectarian terms as a conflict between an Alawi-minority tyranny over a Sunni-majority population. As such, a majority of 86% assigns responsibility to Assad's government and its support through international players while only a miniscule minority of 3% assigns responsibility also to the opposition. Furthermore, 62% name Syrians as the collective victims, whereas 38% single out a specific group, most commonly a specific ethnic or sectarian group.

In the Palestinian case, 80% of respondents interpreted the conflict with Israel through a political lens, viewing it as a territorial and political struggle, rather than one driven by ancient sectarian divisions. Overall, the dominant Palestinian narrative is one of Israeli occupation and the resulting Palestinian losses and collective suffering. However, the distribution of responsibility varies widely within the Palestinian sample. While 43% of respondents placed sole blame on Israel, a majority (56%) also acknowledged some degree of Palestinian responsibility for the ongoing conflict, and 13% assigned responsibility solely to the Palestinian side. A collective sense of Palestinian victimhood also prevailed among Palestinian respondents (82%). In the 18% of cases where specific subgroups were identified as the main victims, these groups were defined by their political marginalization within the broader Palestinian collective—such as Gazans and refugees.

Each sample group thus shows differences not only in their explanations of why the war is taking place, but also in how they identify perpetrators and victims. Perpetrators are defined narrowly as political leaders, or more broadly to include entire populations. In the Ukrainian sample, a majority held both Russia's leaders and its citizens responsible. In contrast, respondents in the Syrian and Palestinian samples predominantly assigned responsibility to political elites. Similarly, victims can be defined broadly, as in the view that 'all people are suffering', or narrowly, as 'a specific group'. In all three samples, the majority of respondents identified the national collective as the main victim group. A minority in each sample singled out particular groups as victims. In doing so, Ukrainian and Palestinian respondents followed a political logic, naming those who became refugees, lived under occupation, or died on the front lines. By contrast, in the Syrian sample, the identification of specific victim groups followed an ethnic or sectarian logic, with both ethnic minorities and the Sunni majority named as primary victims.

Taken together, these patterns suggest that narratives of war function as moral frameworks, that organise experiences of political violence into stories of perpetrators and victims, thereby linking interpretations of trauma to broader understandings of national attachments (Chapter 4) and imagined futures in peace (Chapter 5).

4
National Attachments During Wars

Periods of war and conflict place national identity under particular strain, reshaping individuals' attachment to the nation. Chapter 4 explores national attachments among Ukrainians, Syrians, and Palestinians, using online survey data to map their feelings of belonging and national pride.

National identity refers to the attachment to the nation over sub- or supranational identities. It is an individual's sense of belonging to the nation state, and as such is a fundamental social construct. National identity is never a fixed entity but rather a dynamic cluster of tendencies and values that evolve over time (Abdelal et al. 2009; Bechhofer and McCrone (eds.) 2009; Fukuyama 2018; Malešević 2011b; Lebow 2016; Parekh 1995). As major social, traumatic events, wars and conflicts profoundly affect national belonging. However, in which direction? Do individuals feel more or less attached to their country when it had slipped into a war?

Existing research on the relationship between national attachment and war offers no settled conclusion. A long-standing line of argument, dating back at least to W. G. Sumner's claim that 'the exigencies of war with outsiders are what makes peace inside' (1906, 12), holds that external conflict strengthens national identification by promoting internal cohesion and hostility toward out-groups. This logic has found support in a substantial body of subsequent work (e.g. Sambanis and Lee 2023). However, more recent scholarship has challenged the assumption that warfare automatically generates national solidarity, arguing instead that durable national attachment depends on sustained institutional processes, including ideologization and coercion (Malešević 2011a; Hall and Malešević 2013, 1–7).

International Relations (IR) scholarship similarly suggests that conflict—or the threat of conflict—can shape, reinforce, or transform national identifications (Barnett 1995; Caselli and Coleman 2006; Herbst 1990; Kaufmann 1996a and 1996b; Lake and Rothchild 1996; Levy and Thompson 2010; Tilly 1990; Quillian 1995). Moreover, the effects of war on national attachment vary by conflict type. While interstate wars, particularly those involving external aggression, tend to present a clear and unifying threat to the national group, civil wars often fragment national communities into smaller,

Legacies of War. Kathrin Bachleitner, Oxford University Press. © Kathrin Bachleitner (2026).
DOI: 10.1093/9780198982913.003.0005

competing units (Rotberg 2004; Ayoob 2007; Bates 2008). It therefore follows that different types of war and threat environments are likely to produce distinct effects on individuals' national attachments.

Moreover, studies distinguish between various types of national attachments, such as national identity, nationalism, and patriotism (e.g. Huddy, Del Ponte and Davies 2019). They all interact with war in diverse ways. Patriotism and nationalism in particular have been shown to increase in the face of political violence (Wimmer 2014; Wimmer and Min 2009; Bar-Tal et al. 2014). This is because traumatic experiences tend to strengthen group ties and social bonds among those who have been victimized (Alexander 2012; Vollhardt and Bilewicz 2013). The surge in national and patriotic sentiments described by sociologists describe in social psychology is termed the 'conflict-cohesion hypothesis'. First proposed by Georg Simmel in 1908 (1955), it argues that conflict with an out-group intensifies the cohesion and political centralization of the in-group (Levy and Thompson 2010, 100–101). Equally, Ralph Darendorf (1958, 58) called it 'a general law that human groups react to external pressure by increased internal coherence'.

This chapter presents Ukrainian, Syrian, and Palestinian respondents' feelings of national belonging during times of war. By definition, national identity refers to attachment to one's country over sub- or supra-state identities, and it is commonly approximated through a standard measure used in the World Values Survey: self-assignment to the group identity with which respondents feel they primarily belong.[1] While allowing respondents to choose among different categories reveals whether a sense of national belonging exists and whether it prevails over sub- or supra-state forms of identity, such assignments provide limited insight into the *quality* of that national attachment. To address this dimension, the World Values Survey includes a question that asks respondents to rank their levels of pride in being part of a specific national group.[2] The survey used in this study also included this ranking but went a step further: in order to understand the *sources* of national pride, it asked respondents to elaborate on the reasons why they feel proud to be Ukrainian, Syrian, or Palestinian.

The reasons respondents gave for their national pride were broadly categorized into cultural and political sources. Through qualitative content

[1] See World Values Survey at https://www.worldvaluessurvey.org/WVSContents.jsp (viewed 26 April 2025).

[2] Please note that these measurements are only very crude approximations of the highly complex social phenomenon that is 'national identity'. Furthermore, pride should also not be mistaken as a proxy for national identity. Pride is a variable, changeable, and contextual emotion that conveys contemporary expressions of feelings rather than being a reliable indicator of the long-term stability of national identity. In the context of this study, national pride, and the reasons for it, should thus not be conflated with the stability of national identity/national attachments (See Malešević 2011b, 278–279).

analysis, it emerged that where political reasons prevailed, they were closely tied to the war effort. In such cases, a nationalistic logic appeared to shape respondents' sense of national attachment. By contrast, when cultural reasons were cited as the primary source of pride, respondents seemed to express a more patriotic form of attachment. By definition, nationalism asserts the superiority of one's nation, often at the expense of others, whereas patriotism implies a love for one's country that is not necessarily exclusionary (Viroli 1997; Wimmer 2019). When high levels of national pride are grounded in culture, tradition, and history—rather than in overtly political narratives—these sentiments can therefore be described as patriotic. However, the boundaries between nationalistic and patriotic logics are often blurred in empirical contexts, particularly in translated responses. Against this backdrop, the next section presents respondents' own accounts of why they feel proud to belong to their national in-groups.

National Attachment and Patriotic Pride in the Ukrainian Sample

In the Ukrainian sample, 94% of respondents identified primarily as 'Ukrainian', representing an overwhelming majority. By contrast, only a small number selected a supranational 'European' identity (3%) or substate identities such as 'Crimean Tatar' (2%). Only one respondent identified as 'Russian' (see Figure 4.1).

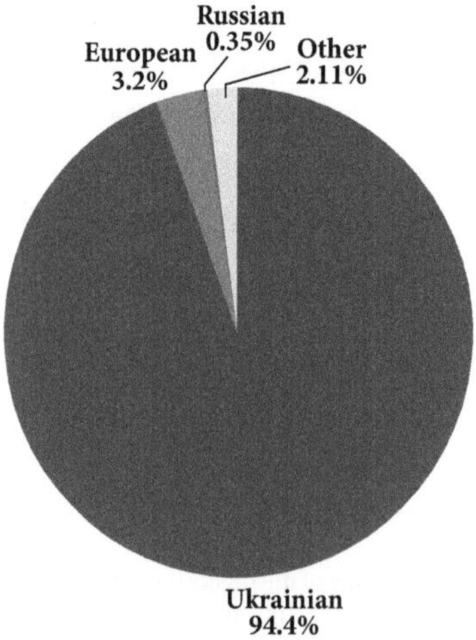

SELF-IDENTIFICATION

●	Ukrainian	269	N = 287
●	European	9	
●	Russian	1	
●	Other	6	

Figure 4.1 Primary Group Identification Among Ukrainian Respondents

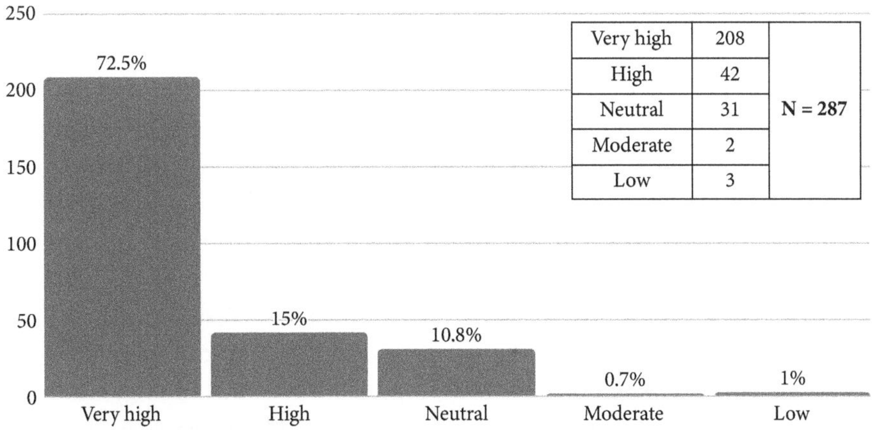

Figure 4.2 National Pride Levels Reported by Ukrainian Respondents

Seventy-two per cent of respondents ranked their patriotic pride as 'very high', and an additional 15% ranked it as 'high'. By contrast, 10.8% selected 'neutral', and two respondents (0.7%) chose 'moderate'. Only 1%—three respondents—rated their patriotic pride as 'low' (see Figure 4.2).

Ukraine: *'A nation of the unyielding'*

Within the Ukrainian sample, 71.3% of respondents explicitly stated that the national resistance against Russia's aggression was the source of their national pride and had significantly strengthened their attachment to the nation. These respondents highlighted the unity and resilience of Ukrainians in the face of invasion, along with the people's bravery, morale, and perseverance in difficult times. For example, a twenty-year-old woman from Crimea, now living in Germany and identifying as 'Ukrainian', stated: *'I appreciate Ukrainian values such as honour and dignity, the pursuit of freedom, and the willingness to sacrifice everything to preserve it. I also admire the high level of cooperation and mutual support among Ukrainians'.*

Similarly, a twenty-three-year-old man from Irpin remarked: *'Ukrainians have always fought—and will continue to fight—for their land, their people, and their independence. I do not know a single nation in the world as brave as ours. That is why I am proud to be Ukrainian'.* A sixty-eight-year-old woman from Chernihiv, currently residing in Ternopil, echoed this view: *'Ukrainians are courageous people who love and defend their country'.*

A sixty-one-year-old woman from Kyiv expressed her pride in the nation's cultural and historical persistence in the face of external aggression: '*We have a rich and glorious history and noble roots. We have an authentic and fascinating culture, language, and many historical landmarks. Many educated and success-oriented people live among us. We value and defend our INDEPENDENCE! We also have an unshakable belief in the bright future of our country—its development and prosperity—despite everything. We are a nation of the unyielding!*'

In the same vein, a forty-year-old woman from Odessa, now living as a refugee in Greece, stated that she was very proud to be Ukrainian:

> Because our nation is unbreakable, we always unite in difficult times. Because we have many heroes who are ready to lay down their lives for the country and for our future. Because this year of war has revealed how cool, indomitable, and beautiful our people are—heroes, soldiers, doctors, volunteers, service workers, and more. Because despite all the difficulties, we try to live on, rebuild, help, restore the country, find strength, and support each other. We are a beautiful, strong nation. Everything will be Ukraine!

Irrespective of gender, level of education, or location, most Ukrainian respondents indicated that the war had strengthened their patriotic pride. Many also contextualized their responses within a longer historical timeframe, praising the heroism, resilience, and unity of Ukrainians over generations. For instance, a twenty-seven-year-old woman from Prymorsk in the Zaporizhzhia region, now living in Italy as a refugee, stated: '*Ukrainians are a strong nation that has endured numerous trials and genocides, yet we have survived and preserved our history, culture, and language*'. Similarly, a thirty-year-old man from Cherkasy, now in Germany, expressed pride because:

> Ukrainians have become synonymous with courage and patriotism. Ukrainians are currently shaping themselves as a full-fledged nation. We are finally leaving behind—not forgetting, but leaving behind—our Soviet past and rejecting the intrusive patronage of the Russian Federation. Russia's policy towards Ukraine for centuries has been purely colonial. Ukrainians refuse to be vassals of an authoritarian regime and are willing to pay the highest price for it. That is why I am proud to be Ukrainian.

A thirty-one-year-old man from Lviv, now living in Poland, similarly expressed very high levels of national pride: '*Because I was born Ukrainian.*

We are a nation that has been oppressed and targeted for destruction for years, but none of our main enemies have succeeded. Despite this, we continue to live and raise children. It is just unfortunate that we were not able to realize our potential in our homeland before the full-scale invasion.

Taken together, these responses illustrate that Ukraine's collective war effort in the face of both historical and current oppression has augmented national unity and attachment. Respondents also frequently linked their feelings of pride to the ideal of freedom. For example, a fifty-year-old woman from Dnipro stated that she was very proud to be Ukrainian '*because we are a free nation!*' Similarly, a sixty-year-old man from Irpin expressed very high pride, explaining, '*I feel like a free person*'.

Interestingly, high levels of pride were also found among those respondents who did not primarily identify as Ukrainian and even among those who explicitly rejected the concept of patriotism. A twenty-nine-year-old woman from Simferopol (Aqmescit), living in Crimea and self-identifying as 'Crimean Tatar', ranked her pride in being Ukrainian as 'neutral' but wrote: '*I, like any person, have a hierarchy of identities. For me, my Crimean Tatar (ethnic) identity comes first. Therefore, it is difficult for me to answer. However, if you look at this issue from the point of view of citizenship, then for me there is certainly something to be proud of. I am proud of the resistance and, first of all, the freedom and will of the Ukrainian people*'.

An eighteen-year-old woman from Kharkiv, now living in Germany as a refugee, explicitly rejected patriotism yet ranked her national pride highly: '*I consider belonging to a nationality not a reason for pride, because it is determined by the fact of birth and not acquired as a result of a conscious choice. However, I still ranked myself highly on the scale because the choices made by the people to whom I belong are worthy of respect and memory*'.

Adopting a similar reasoning, a thirty-two-year-old man from Kamianets-Podilskyi, who self-identified as 'Ukrainian', stated: '*I do not think that nationality or citizenship should play an important role in the assessment of a person, but the ability to consolidate in a difficult period and fight for one's home is an outstanding quality of Ukrainians. It had a rather strong impact on the whole world. The security of at least Europe now depends on it*'.

From the words of many Ukrainian respondents, it is clear that some have become more patriotic and nationalistic since the annexation of Crimea in 2014, and especially since the outbreak of full-scale war in February 2022. This trend is explicitly stated, for instance, in the response of a twenty-one-year-old woman from Vinnytsia, now living as a refugee in Warsaw: '*The increase in pride for my country has occurred in tandem with the rise of nationalism among Ukrainians against the backdrop of the full-scale invasion*'.

Similarly, a thirty-one-year-old woman from Kharkiv, now living in Vienna, said:

> Until 2014, I was more cosmopolitan. Although still a patriot, I considered nationalism to be a harmful vestige. I have always loved my country, but I thought that being proud of a national flag was a rather pointless endeavour. However, with the beginning of the war—especially after the full-scale invasion—I realized what it means to be proud of being Ukrainian. Even during the Maidan, when I saw the enormous drive and thirst for freedom my people have, I could not help but be proud of them.

Some respondents also reflected on how Russian political and cultural influence had previously suppressed their sense of Ukrainian national pride. A twenty-four-year-old man from Kyiv, now living in Poland, and who identified himself as possessing 'very high' levels of patriotic pride, explained:

> Before the full-scale invasion, I was less proud to be Ukrainian—not because I wanted to be someone else, but because, like most Ukrainians, I had a sense of inferiority, mostly due to Russian propaganda actively disseminated through TV series since the early 2000s. But now I am proud to be Ukrainian, and proud of our history, our thirst for freedom, which we have fought for—and continue to fight for.

Another man, a twenty-nine-year-old from Nikopol currently based in Lviv, identified himself as 'neutral' on the pride scale. He explained that he once felt a nationalistic attachment to Russia but is now moving towards a more cautious patriotism for Ukraine:

> Before the large-scale invasion, I considered myself apolitical, spoke Russian, and consumed Russian media, literature, and music. With the onset of the war, my views gradually began to change (I am still in the process of separation). I am proud that my people were able to unite against a fierce enemy and demonstrate military and civilian victories. Choosing between the 'Russian world', with blackouts and thousands of fallen soldiers and civilians, or life in a declining economy—I choose the latter. However, my impression is spoiled by individuals and groups who seem not to have learned anything from the war. I refer here to those in power—from minor officials and Soviet-school military officers to the higher echelons of government. When someone in power acts responsibly, it is perceived positively, but it is also expected. When they deceive, engage in corruption, or abuse their authority, it delivers a strong blow to my faith in the country's future. Therefore, my assessment of national pride is somewhere in the middle. I cannot predict the future, so I do not know whether I will ultimately feel proud.

While most Ukrainian respondents reported increased national attachment as a result of the war, a small number indicated the opposite. For example, a twenty-four-year-old man from Zaporizhzhia, now living in Kyiv, expressed little pride in being Ukrainian. He attributed this to the delayed recognition among Ukrainians that Russia is an aggressor, and to how closely Ukraine, in his view, still resembles Russia:

> I am proud that the majority of society has finally realized who the enemy is, after nine years of war, and that this enemy has failed to achieve what it planned. However, I am not proud because the war is also the result of the actions of Ukrainians during the years of independence. They voted for pro-Russian parties and presidents, considered Russians as brothers, were against NATO, and looted the country. Around 60–70% of the population considered Putin a good president, and 80% were either against NATO or undecided. Even now, during the war, the country continues to be plundered—and it is shameful. And this is the fault of the entire Ukrainian people, who refuse to acknowledge it, just as Russians refuse to acknowledge their own guilt for the war. I am also ashamed that society has started supporting human rights abuses, such as restrictions on movement and discrimination based on gender—which, even during a state of emergency, according to the Ukrainian constitution, should not exist. In this regard, Ukraine has turned into North Korea, and society seems to be fine with it. I understand the radicalization regarding the Russian language, but it often reaches absurdity. I am ashamed that 50–70% of the population (judging by Telegram channel reactions to a petition for same-sex marriage) are homophobic. There are organizations that beat people for their sexual orientation, and others that do so based on appearance. In this regard, Ukraine resembles Russia and other CIS countries.

A twenty-seven-year-old man from Petrivka in the Dnipropetrovska region, now living in Chernivtsi, similarly rejected all forms of national identity, instead identifying as a *'human of the universe'*. He placed himself at the lowest level on the pride scale and explained: *'I am not proud because of the government's lies to the population about there not being a war, even though they knew it would happen. Criminal mobilization, border closures, unemployment, censorship, corruption, theft, and mistreatment of our soldiers. I would like to forget this country like a terrible dream forever'.* This respondent reported that his patriotic pride had diminished after being forcibly drafted into the army.

Likewise, a thirty-year-old man from Mariupol, now living in Wrocław, Poland, rated his pride as 'neutral', noting both reasons for pride and disappointment: *'I am proud because we are on the path of democratization. I am not proud because the state has some totalitarian aspects in its constitution—for instance, forced mobilization, and many others*. He too cited Ukraine's mobilization policies as the main factor tempering his feelings of national pride.

Interestingly, most respondents who reported decreased patriotic sentiment were young men who had experienced gender-based discrimination under Ukraine's mobilization rules. Perhaps unsurprisingly, the only respondent who self-identified as 'Russian' also expressed declining attachment to Ukraine. This thirty-four-year-old woman from Kharkiv, now living in the UK, rated her patriotic pride as 'neutral' and commented: '*Ethnic Ukrainians kill and discriminate against people who belong to the nationality "Ukrainian" but have a different ethnicity*'.

While the majority of respondents grounded their patriotic pride in the current war effort and resistance to Russian aggression, 21% attributed their pride primarily to cultural factors, such as Ukraine's traditions, cuisine, language, history, and people. For instance,, a thirty-one-year-old woman from Kharkiv explained her 'very high' pride as follows: '*We are a friendly nation. We have a very bright culture, and creative, beautiful people*'. A thirty-two-year-old man from Kyiv, now living in London, similarly stated that she is proud because of '*Ukraine's culture, history, language, architecture, and unique characteristics*'.

A sixty-two-year-old woman from Kharkiv, now in the UK, described her pride in purely cultural terms: '*Ukraine is my homeland, my identity, my culture, the land where my ancestors lived and live, my native language. It is an ancient Christian beloved land. The key words here are native, familiar, mine, beloved, energizing, and inspiring. Ukraine is my resource. I feel fulfilled and happy here*'.

A thirty-four-year-old man from Kyiv, now living in Prague, explained his pride in similar terms: '*I love Ukraine because of our amazing technologies that are not available in Europe. Lots of delicious food, the most talented people, the most beautiful language*'. Finally, a fifty-year-old woman from Donetsk, now living in Austria, stated simply: '*Ukraine is beautiful, fashionable, tasty, and intelligent*'.

To summarize the findings from the Ukrainian sample, a majority of respondents (71%) attributed their high levels of patriotic pride to the ongoing effort to repel Russia's aggression (see Figure 4.3). These individuals frequently depicted Ukraine as a 'nation of the unyielding' in the face of both current and historical challenges to its national distinctiveness. The war also appears to have heightened the pride of a small number of respondents who rejected the

War	204	
Cultural	61	
Mixed	7	
Not patriotic	14	
Total	**N = 286**	

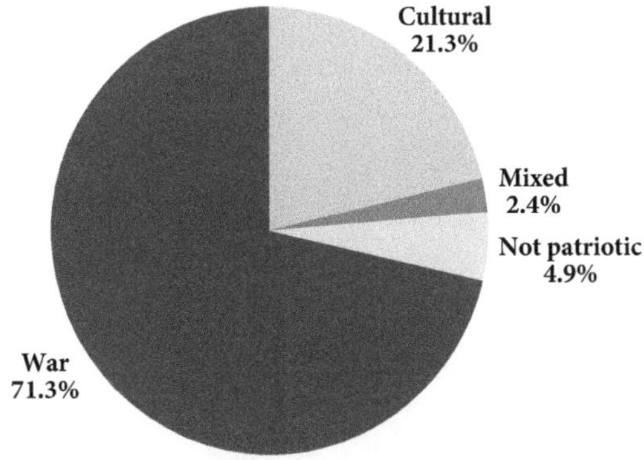

Figure 4.3 Stated Reasons for National Pride Among Ukrainian Respondents

notion of patriotic pride—whether on cosmopolitan grounds or due to their identification with a subnational ethnic group. Only 4.9% of respondents stated explicitly that they did not consider themselves patriotic.

Overall, the politics surrounding the current war with Russia seem to have deepened national attachment among Ukrainians. Yet among the small number of respondents who expressed low levels of patriotic pride, the war was also cited as the main reason for their detachment. These respondents, in particular, voiced opposition to the Ukrainian war effort, most commonly in relation to the policy of forced mobilization of young men. In this context, respondents' levels of patriotic pride appear closely tied to their political stance on the war: for some, the war augments their pride; for others, it diminishes it.

While the majority rooted their pride in the war effort, a significant proportion—21%—cited purely cultural reasons, referring to Ukraine's history, language, traditions, or people.

National Attachment and Patriotic Pride in the Syrian Sample

In the Syrian sample, 53% of respondents identified with the national category 'Syrian'. Among the remainder, 19% identified as 'Arab' (a supranational identity), 14.5% as 'Kurdish', and 8.5% as 'Muslim' (Sunni). Only a small number of respondents chose other sectarian identities, such as 'Christian' (0.8%) and 'Alawi' (one respondent) (see Figure 4.4).

Furthermore, in the Syrian sample, 59.5% of respondents ranked their levels of patriotic pride as 'very high', while 7.1% of respondents selected 'high', and 9.5% selected 'neutral'. Among the remaining respondents, 4.1% selected

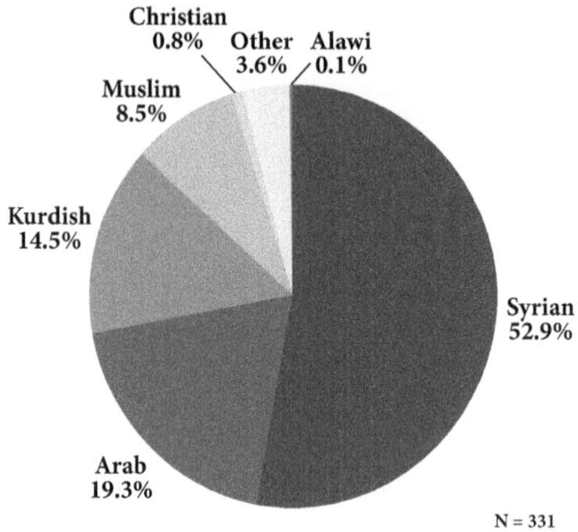

SELF-IDENTIFICATION

●	Syrian	175
●	Arab	64
●	Kurdish	48
●	Muslim	28
●	Christian	3
●	Alawi	1
●	Other	12

Figure 4.4 Primary Group Identification Among Syrian Respondents

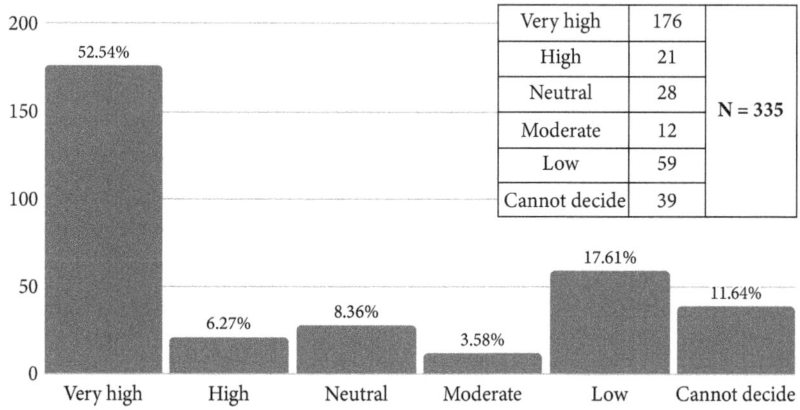

Very high	176	N = 335
High	21	
Neutral	28	
Moderate	12	
Low	59	
Cannot decide	39	

Figure 4.5 National Pride Levels Reported by Syrian Respondents

'moderate', 19.9% selected 'low', and 11.6% stated that they 'cannot decide' (see Figure 4.5).

Syria: *'The Cradle of Civilizations'*

In the Syrian sample, 18.5% of respondents indicated that they are proud to be Syrian for political reasons, specifically referencing the civil war, with most in this group expressing support for Syrians' resistance to the Assad regime. However, 33.8% explicitly stated that the politics of their country—and the civil war that has resulted from it—make them 'not proud' to be Syrian. Moreover, cultural reasons represent the largest source of national pride in the Syrian sample (40.4%) (See Figure 4.6). These responses often refer to Syria as the 'cradle of civilizations', highlighting its ancient historical achievements and its pluralistic mix of religions and ethnicities. Additionally, 7.3% of respondents explicitly rejected the idea of patriotism and nationalism.

The 18.5% of Syrian respondents who attributed their pride to the ongoing war effort all ranked themselves as 'very proud' to be Syrian. Most of these responses cited political resistance to the Assad regime—specifically the collective effort, determination, and sacrifice of the opposition—as the source of their pride. In most cases, this pride was grounded in the belief that Syria and its people are distinct from the Assad regime. Such responses frequently conveyed strong national attachment, describing Syria as their 'motherland' or 'homeland' and emphasizing national unity.

For example, a sixty-five-year-old man from Rif Dimashq, who fled to Lebanon and self-identifies as 'Sunni', explained:

> It is the land where I was born and raised. I belong to its people and its civilization, which is deeply rooted in history. Syrians are naturally hard-working, loyal to their friends, and rich in values and emotions. That is why I am proud to belong to Syria. Even though I reject the regime and those who have attacked our homes and country—as well as the terrorist groups on the other side that have violated Syrian blood, territory, and resources—I still believe Syria will recover. It will triumph over both sides and return to being an oasis of security and peace in the region and the world.

For these respondents, pride in being Syrian stemmed from opposition to the Assad regime and hope for a future Syria. A twenty-nine-year-old man

Cultural	61	●
Not proud because of politics	51	●
War	28	●
Against nationalism as an idea	11	●
Total	**N = 151**	

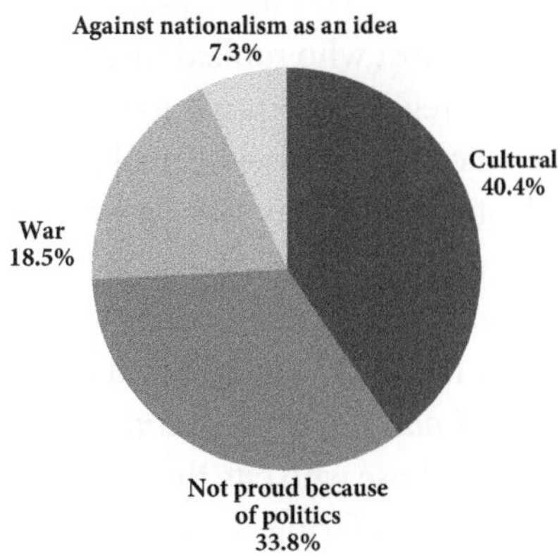

Figure 4.6 Stated Reasons for National Pride Among Syrian Respondents

from Hasakah, now a refugee in Turkey who self-identifies as 'Syrian', stated: *'I am proud because many people from my country have sacrificed their lives to gain freedom and rights'.*

However, on the other side of the political divide, a small number of respondents who supported the Assad regime also expressed patriotic pride due to the war effort. As a twenty-seven-year-old man from Homs, currently living in Saudi Arabia and identifying as 'Alawi', put it: *'Everything happening in my country—the ongoing resistance and the victory to come—makes me proud!'*

In the Syrian case, then, the war has served as a source of patriotic pride for both supporters of the regime and of the opposition.

Moreover, when respondents cited the war as a reason for pride, they often referenced the collective suffering and sacrifice of the Syrian people. A forty-one-year-old woman from Rif Dimashq, now a refugee in Germany and identifying as 'Syrian', stated: *'No one in the modern world has suffered more than the Syrian people'.*

Some respondents also highlighted the discrimination experienced as a result of displacement. A thirty-five-year-old man from Daraa, now in Gaziantep, Turkey, who identifies as 'Syrian', reflected: *'I am deeply proud because of the humiliation we faced in asylum countries. If I'd been asked this question before the revolution, I would have said, "I don't care"—because I believed all people were equal. But now I understand things differently'.* Shared suffering, particularly in exile, appears to have generated a renewed sense of national solidarity among refugee communities (Benhabib, Shapiro, and Petranovic 2007; Betts and Loescher 2010, 18; Kastoryano 2007, 163–168; Van Hear and Cohen 2017).

While 18.5% of respondents grounded their national pride in the war effort, nearly twice that number (33.8%) explicitly stated that the outbreak of civil war and the country's politics had weakened their sense of national attachment. Most who ranked themselves as 'not proud' did so in opposition to the Assad regime, pointing to the chaos, brutality, and injustice wrought by the conflict. A fifty-nine-year-old man from Qamishli, now living in Norway and identifying as 'Kurdish', described it as *'the inhumanity that happens there every day'*.

To provide further examples, a forty-nine-year-old woman from Afrin who identified as 'Kurdish' stated that she is not proud to be Syrian because of the *'moral and political decline, brutality, and barbarism of the Syrian conflict'*. Unlike those who cite the war as a source of national pride, these respondents view the suffering it has inflicted as a cause of disillusionment rather than unity. As one thirty-three-year-old woman from Damascus, now a refugee in the United Kingdom who self-identifies as 'Arab', put it: *'Syrians have suffered deeply as a nation'.*

The erosion of national attachment resulting from the conduct of the Syrian civil war is also evident in the response of a thirty-six-year-old man from Afrin, now an internally displaced person in Qamishli and identifying as 'Kurdish'. He explained: *'I am not proud because Syrians have failed to unite for the good of the homeland; instead, they have been manipulated by external powers'.* Other respondents moved beyond the immediate context of the war and pointed to Syria's broader authoritarian legacy. For instance, a thirty-year-old man from Kobani, also living as an IDP in Qamishli and identifying as 'Arab', cited *'oppression and injustice'* as reasons for his lack of national pride. Similarly, a forty-year-old man from Qamishli who self-identifies as 'Kurdish' stated: *'I am not proud because I stayed silent for forty years under an unjust regime. And now, the revolution has turned into an Islamic movement'.*

All the respondents in this category expressed low levels of national pride, tying their sentiments to the country's political situation. Some have gone so far as to reject Syrian national belonging altogether. A sixty-one-year-old man from Rif Dimashq, now a refugee in Turkey, stated: *'Syria is not a democratic country and does not respect human rights. I no longer want to belong to it'.* Others voiced disillusionment not only with the political regime but with Syrian society itself. As a forty-six-year-old man from Homs, currently in Turkey, explained: *'Syrian society lacks manners and values. Its people behave like mercenaries'.* A seventy-five-year-old man from Afrin, now living in Germany and identifying as 'Kurdish', put it more bluntly: *'Syrians change their opinions quickly and sell themselves to the highest bidder'.*

Most respondents who grounded their low levels of national pride in the current political situation did not identify as 'Syrian'. Rather, they aligned with substate identities—most commonly 'Kurdish'—or supra-state identities such as 'Arab', or they stated explicitly that they did not feel they belonged to any group. This may be due to the way nationalism in Syria has become closely associated with the Assad regime over the past decades (Aldoughli 2022; Chevee 2022). In particular, most respondents who identified as 'Kurdish' stated that they were not proud to be Syrian because their ethnic identity is not recognized. Many also referred to their experiences of persecution by the Syrian state. As a sixty-one-year-old Kurdish man from Qamishli, now living in Dubai, explained: *'I am not proud to be Syrian because I have faced persecution there since birth'.*

In sum, political opposition to the Syrian government and ethnic discrimination under the regime have severely weakened many individuals' attachment to the Syrian national identity.

While the political situation alienated 34% of respondents from national attachment, cultural factors emerged as the most significant source of pride in the Syrian sample. A total of 40.4% of respondents reported being 'very proud' to be Syrian because of Syria's ancient history, its civilizational achievements, and its status as the 'cradle of civilizations'. These responses often referenced developments that pre-date the modern Syrian state and highlighted a long-standing tradition of religious and ethnic pluralism. As one sixty-four-year-old woman from Damascus, now living in the US and self-identifying as 'Sunni', expressed: '*We are the ones who built civilization; our history is ten thousand years old*'.

A thirty-one-year-old man from Afrin, identifying as 'Kurdish' and now living in Europe as a refugee, stated: '*I am proud because Syria includes many ethnicities, nationalities, and moral values. It is a blend of nations*'. Other respondents referred to Syria's contribution to Islamic civilization. For instance, a fifty-one-year-old man from Daraa, now living in Germany and self-identifying as 'Arab', noted that his pride stems '*from the fact that I am the son of an Islamic culture with a long history, and from Syria's legacy, which spans thousands of years. This creates a unique diversity in the world*'. Similarly, a fifty-three-year-old man from Daraa, living in Qatar, proclaimed: '*Syria is an Islamic and Arab country, with a long tradition of civilization and great cultural diversity*'.

Several respondents used Syria's historical depth to differentiate between the nation and its current regime. As a thirty-five-year-old man from Homs, now living in Lebanon, put it: '*I love my country and cherish its history before Assad*'. A sixty-seven-year-old man from rural Hama, now residing in Australia and identifying as 'Syrian Arab', similarly asserted: '*Syria is the land of civilization, no matter what unworthy rulers have done*'.

While many of those proud for cultural reasons referred to Syria's ancient civilizational history, others highlighted qualities of the Syrian people themselves. Respondents described Syrians as '*hard-working*' (response from a forty-nine-year-old man from Daraa, now in Turkey, self-identifying as 'Arab') and as '*kind and authentic*' (response from a thirty-two-year-old woman from Rif Dimashq, now in Lebanon, self-identifying as 'Arab').

Meanwhile, 7.3% of Syrian respondents explicitly rejected the concepts of patriotism and nationalism. These individuals often expressed cosmopolitan or individualist values, suggesting that pride should be based on personal achievements rather than birthplace. A fifty-year-old man from Damascus, now in Turkey, said: '*Pride comes from personal achievements, not from where you happen to be born*'. Similarly, a twenty-seven-year-old woman from Daraa, now an internally displaced person, wrote: '*It makes no sense to be proud of things I didn't choose but was simply born into*'. A sixty-five-year-old man from Al-Hasakeh, now living in Germany and identifying as 'Kurdish', echoed this sentiment: '*If anything, I'm proud to be a human being—regardless of my geographic affiliation*'.

Some respondents in this category explained that the conflict itself led them to disidentify with Syria altogether. In their view, the notion of a 'Syrian people' is a recent construct that masks older, more meaningful identities. A twenty-year-old man from Raqqa, now in Germany and identifying with his local 'Euphrates' identity, stated: '*There is no such thing as a Syrian people. Many associate Syrian identity with Arabism, which itself denies the existence of a distinct Syrian identity. To be Syrian would mean belonging to the land and the people who make up Syria, but Syria was created by a colonial pact without the people's consent. My regional identity was erased as a result*'. Similarly, a thirty-two-year-old man from Damascus, currently living in Idlib and identifying as 'Sunni' (a supranational religious identity), remarked: '*The idea of belonging to Syria is less than a hundred years old. I belong to a land that is thousands of years old*'.

These responses suggest that, in the Syrian context, nationalism is rejected on both post-national cosmopolitan grounds and on supranational or anti-colonial grounds. Respondents who opposed national identity either adopted an individualistic outlook or rejected nationalism as a Western imposition ill-suited to Syria's complex social fabric and layered identities—particularly Arab and Sunni Muslim affiliations.

To summarize the findings from the Syrian sample, national pride primarily stems from cultural, historical, and personal attachments rather than contemporary politics. Syria's identity as the 'cradle of civilization' and as a place of pluralistic coexistence appears to generate a strong emotional bond among many respondents. This sense of pride is often rooted in a deep historical narrative that predates both the modern Syrian state and the Assad regime. For others, pride comes from a sense of community, describing Syrians as '*kind, authentic, and hard-working*'.

However, for a substantial number of respondents, the ongoing conflict and political oppression have dampened national pride. Many of those who have grown disillusioned point to the brutality of the civil war, the repressive nature of the Assad regime, and feelings of exclusion or persecution. In contrast, the few respondents who drew pride from the war effort generally did so in connection with their opposition to the regime and their belief in the resilience and sacrifice of the Syrian people. Finally, some respondents rejected national identity altogether—either on cosmopolitan and universalist grounds, or in favour of more resonant sub- or supra-state identities such as Arab ethnicity or the Sunnah religion.

National Attachment and Patriotic Pride in the Palestinian Sample

In the Palestinian sample, 75% of respondents identified primarily as 'Palestinian'. A further 11% identified as 'Muslim', 8% with their status as a 'Palestinian refugee', and 3% with their 'Arab' ethnicity (see Figure 4.7).

Furthermore, 70% of respondents ranked their levels of patriotic pride as 'very high', 12.5% selected 'high', and 8.6% described themselves as 'neutral' (see Figure 4.8). On the lower end of the spectrum, 3.8% identified as being 'moderately' proud to be Palestinian, while 4.8% reported low levels of pride.

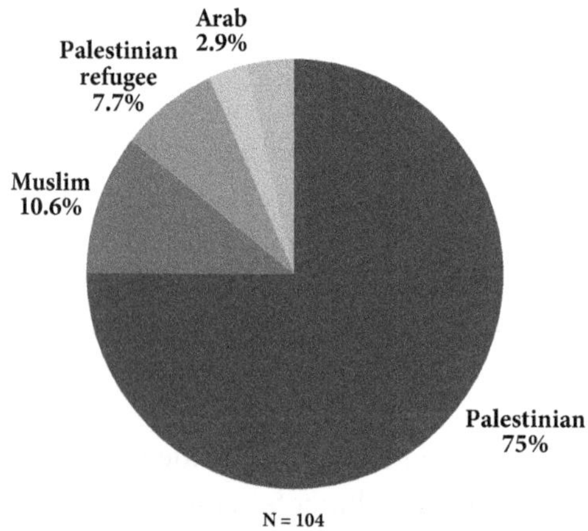

SELF-IDENTIFICATION

●	Palestinian	78
●	Muslim	11
●	Palestinian refugee	8
●	Arab	3
●	I don't belong	4

Figure 4.7 Primary Group Identification Among Palestinian Respondents

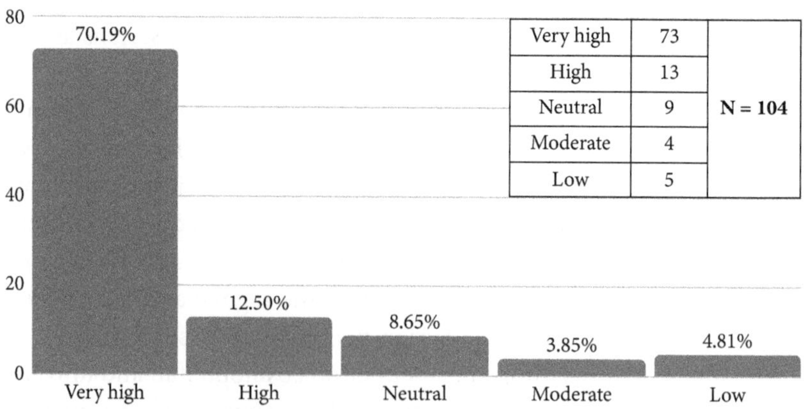

Figure 4.8 National Pride Levels Reported by Palestinian Respondents

Notably, four respondents in the Palestinian sample stated that they did not feel they belonged at all—an answer not found in the other two samples.

Palestine: '*My Territory, My Country, and the Land of My Ancestors*'

In the Palestinian sample, respondents predominantly cited political and cultural factors as the basis for their national pride. Of these, 44.2% identified the 'Palestinian struggle and resistance' as their principal source of patriotic pride, while 41.3% referred to cultural elements, including Palestine's historical and religious significance and its distinctive traditions (see Figure 4.9). Meanwhile, 12.5% of respondents reported they were not proud to be Palestinian, primarily due to political factors. These included factional fragmentation, corruption, and dissatisfaction with the Palestinian Authority's handling of the Israeli–Palestinian conflict. In this context, politics and protracted conflict appear as both sources of national pride and drivers of detachment.

However, for the majority of respondents (44.2%), the 'Palestinian struggle and resistance' is a source of pride. In their view, being Palestinian is not only a matter of identity—of group belonging—but also a political cause. A thirty-six-year-old man from Gaza who identifies as a 'Palestinian refugee' explained: '*Palestine is my identity, my Arab identity, and my cause. I belong to it, and it belongs to me*'. Similarly, a twenty-one-year-old woman from Jenin, currently living in Birzeit and strongly identifying as 'Palestinian', stated: '*I feel proud because we have a cause. The story is the goal, and the goal represents me*'. A thirty-seven-year-old man from Beersheba, now living in Gaza, echoed this connection between pride and political cause: '*I fight for an honourable*

Resistance	46	
Cultural	43	
Not proud because of politics	13	
Against the idea of nationalism	2	
Total	N = 104	

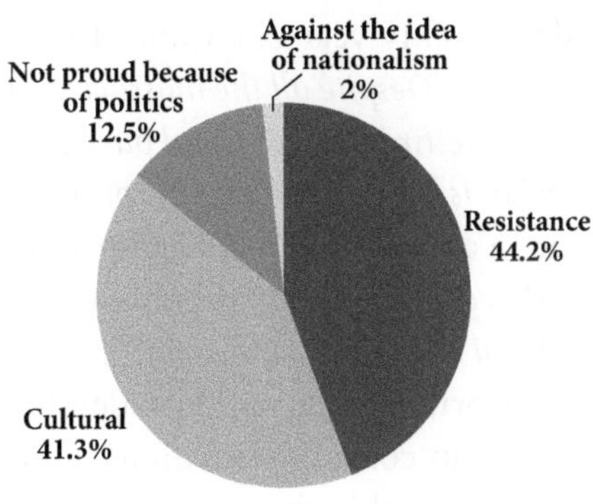

Figure 4.9 Stated Reasons for National Pride Among Palestinian Respondents

cause—the return of my country and the recovery of my rights to the home that was stolen from me'. Likewise, a forty-two-year-old man from Beersheba, now in Nuseirat refugee camp and identifying as 'Muslim', remarked: '*We are a militant people who will never give up our rights. We surpass the occupation in education and morals. We excel in finding solutions to all the problems caused by the occupation—siege, destruction, killing, and displacement*'.

Many respondents emphasized pride in their people's continued resistance to the Israeli occupation, highlighting traits such as Palestinian 'patience and endurance' (a twenty-one-year-old woman from Ramallah) and 'steadfastness' (a thirty-seven-year-old man from Gaza). A forty-year-old man from Lod, now in Gaza, wrote: '*Because I am a steadfast, determined, and patient Palestinian under occupation. That is why I am proud to be Palestinian, and proud of my Arab identity*'. Others framed the struggle in religious terms, invoking jihad or the defence of Islam. As one twenty-three-year-old man from Ramallah put it: '*I'm proud because we defend Islam*'.

In addition to emphasizing ongoing resistance, many respondents pointed to their association with a 'just cause' as a source of pride. In the words of a twenty-three-year-old man from Ramallah: '*Belonging to a just cause is what makes me proud*'. A thirty-seven-year-old woman from Nablus, now living in Ramallah, explained: '*I'm part of a just cause. I belong to a beautiful country—socially, geographically, and historically*'. A forty-year-old man from Sarafand al-Kharab, now in Gaza and identifying as a 'Palestinian refugee', remarked: '*I'm proud because I am Palestinian, because I have a just cause. In my view, Palestine is the best of all lands—the cradle of religions and the home of prophets*'. A twenty-six-year-old woman from al-Majdal, now in Gaza, similarly stated: '*I am the bearer of a cause, a principle, and a land. These are constants I will never give up*'.

However, the sustained effort to realize this cause has also given rise to a collective sense of victimhood, which, in turn, has become a source of pride. A twenty-year-old woman from al-Ramla, currently living in Ramallah, expressed: '*Despite all the massacres, victimhood, and discrimination inflicted on the Palestinian people and our land, we are still standing, still trying to protect what is left. That makes me proud!*' An eighteen-year-old man from Hatta, living in al-Arroub refugee camp and identifying as 'Muslim', remarked: '*I'm proud because Palestinians have endured every catastrophe and still serve as the shield defending Jerusalem*'.

As theories linking national identity and external threats suggest, threats to identity can consolidate national attachment through shared sacrifice. This idea is reflected in the words of a twenty-two-year-old woman from Ramallah: '*It is a country of sacrifice and eternal struggle—a land of patience. It supports its people despite the conditions they live in, and its identity is rooted in the land. It is mentioned in the Qur'an, the land of the Mi'raj, and the site of the Day of Resurrection*'. A twenty-three-year-old man from Anabta similarly stated: '*When your identity is under threat of occupation and theft, you must preserve it—and be proud of it!*' A twenty-two-year-old woman from Rawabi, currently living in New York, explained her pride as deriving from '*the strength and optimism of our people, despite all difficulties*'. For some, this identification becomes a source of action. As one fifty-seven-year-old man from al-Bireh, now in New York, declared: '*We must be proud of who we are. A person who is proud of their background is better able to change their reality and shape their future. As Palestinians, we have preserved our identity and the hope for justice, despite being displaced by powerful forces backed by the world's leading country—the US*'.

Given the nature of the ongoing conflict with Israel, nationalism and patriotism are inseparable from the Palestinian struggle for land and rights. This is evident particularly among respondents who rooted their pride in ancestry. A sixty-four-year-old man from Ashkelon, now in Gaza, stated: '*I'm proud because this is my homeland, my country, the land of my ancestors*'. A thirty-seven-year-old man from Gaza, identifying as a 'Palestinian refugee' from Jaffa, explained: '*I am proud because it is my country, my mother—its capital is Jerusalem!*' A thirty-eight-year-old man from the diaspora, living in Brooklyn, New York, proclaimed: '*It's who I am—my culture, my blood, my history!*' A twenty-two-year-old man from Nablus, identifying as 'Muslim', summed it up: '*The grandeur of the name "Palestinian" alone is enough to make me proud*'.

Interestingly, in the Palestinian case, historical struggle is a clear source of pride for the majority of respondents, whereas current politics often is not. As a forty-year-old man from Beit Jala who identifies as 'Palestinian' put it: '*The main reason I'm proud is that we have managed to survive despite all attempts to obliterate our identity and remove us from the map. The reason I'm not proud is the corruption among those in power, the suppression of freedom, authoritarianism, internal fighting, and the failure to develop a national strategy that can unite our energy and free us from occupation*'.

A similar perspective was shared by a nineteen-year-old woman from Hebron, now living in the Netherlands:

> I'm proud to be Palestinian because of our rich history of resistance and heroism. However, I'm not proud of the mistakes we've made throughout history. Politically, I'm not proud of our current representation, the Oslo Accords, or our constitution, which simply copies the Jordanian one and isn't progressive—it's anti-feminist and not secular. Palestinians value education, and I'm proud of that,

> but our government is full of old men. Palestine needs its youth, and we should move beyond emotional judgements and heroic rhetoric, and focus on ourselves as a nation. Israel is a colonialist regime, but the world doesn't seem to understand this. We should stop relying on others to save us. Israel is not going anywhere, and we must accept that. Not all of our problems come from the occupation; many arise from our own backward decisions. When I say I'm proud to be Palestinian, I mean that I'm proud of the person I've become because of my experiences living there.

These respondents express a mix of pride rooted in historical resilience, and frustration with present-day Palestinian politics. In total, 12.5% of respondents explicitly stated that they are not proud to be Palestinian due to political dysfunction, particularly pointing to factional divisions, corruption, and mismanagement by the Palestinian Authority. For example, a twenty-eight-year-old woman from Tulkarem noted: *'I'm not proud of the divisions, favouritism, corruption, and lack of respect for citizens'.*

While most of these respondents criticized political elites, some—especially younger participants—also expressed disillusionment with broader societal values. A twenty-five-year-old man from Ramallah, now living in Birzeit, commented: *'I don't feel like I belong because of the political and economic situation, the culture, and the consequences of the failed societal norms within which we live'.* Similarly, a twenty-two-year-old woman from Ramallah, identifying as 'Arab', stated: *'I'm not very proud to be Palestinian because I'm ashamed of my people and their actions. I love Palestine, but I hope its people will do more to preserve and develop their homeland'.* A twenty-two-year-old woman from Nablus, also in Ramallah, expressed concern about *'the mindset and culture of most Palestinian citizens'.*

Others cited lack of freedom or mobility as a source of diminished national pride. A sixty-seven-year-old man from Bethlehem referred to *'the lack of freedom that comes with a Palestinian passport'*, while a thirty-six-year-old man from Nablus, now in Germany, stated: *'Being Palestinian has held me back my whole life'.* For Palestinians living inside Israel, expressions of national pride were often fraught. As one twenty-seven-year-old woman from Jaffa said: *'It's hard to feel proud as a Palestinian in Israel'.*

Several respondents in this category also described the national struggle not as a source of empowerment, but as a burden. A forty-two-year-old woman from Burqa, now in Nuseirat Camp, said: *'I'm not proud—it's a lifetime of suffering'.* A thirty-three-year-old man from Gaza, identifying as 'Muslim', noted: *'I'm not proud because I don't have a free and independent country'.* In the Palestinian sample—as with some of the Syrian respondents—struggle and sacrifice fostered pride for many, but led to disillusionment for others.

Alongside political responses, 41% of Palestinian respondents cited culture as the primary source of national pride. Given historic Palestine's role as the cradle of the three monotheistic religions, it is unsurprising that these answers frequently included a strong religious element. Several respondents referenced Palestine as *'the land of the Prophet Mohammed'* (a thirty-seven-year-old man from Gaza) or as *'a sacred Arab land'* (a thirty-two-year-old woman from Yazur, now in Gaza, identifying as a 'Palestinian refugee'). Christian Palestinians pointed to Bethlehem's religious significance. A forty-one-year-old woman from Bethlehem, now in Ramallah, explained: *'I'm proud of my country because it's a holy place—Jesus Christ was born there, lived there, and walked on its soil'.* From an Islamic perspective, a thirty-eight-year-old man from Gaza described his pride in the *'Palestinian identity because Palestine is the land of the Prophets, the site of the Day of Resurrection, and home to the greatest mosque, Al Aqsa'.*

Some respondents grounded their cultural pride in everyday practices. A thirty-five-year-old Palestinian woman from Amman, now living in New York, stated: *'Being Palestinian is my life—it's the food I eat, the clothes I wear, the work I do. It's the basis of my entire identity'.* Others cited characteristics of the Palestinian people. A forty-year-old man from Yibna, now in Gaza, expressed pride in Palestinians' *'masculinity, generosity, and eloquence'.* Likewise, a sixty-two-year-old woman from Gaza, now living in the United States, stated: *'I'm proud to be Palestinian! It's the land of my father and grandfather, and my birthplace is home to intelligent, striving, patient, and dignified people'.*

While history, tradition, and religion were consistently sources of pride, some respondents in this category still qualified their sentiments due to current events. A twenty-two-year-old woman from Hebron, identifying as a 'Palestinian refugee', explained: *'I'm proud because of our history and because Palestine is a sacred land. But I'm not proud of the current situation, the conflict, or the political leaders and their interests'.* For many, even cultural pride could not be entirely disentangled from contemporary politics.

Among the few who rejected nationalism outright (2%), politics again featured prominently. A twenty-one-year-old man from Ramallah, identifying as a 'Palestinian refugee' from Jaffa, explained: *'My answer is neutral. I can't be proud of something I didn't choose. But I'm not ashamed of being Palestinian, and I'm not trying to erase or change my identity to keep the peace. I just can't take pride in it'.*

In sum, the Palestinian sample reveals consistently high levels of national pride (85%). Among these, 44% cited political reasons—chiefly the ongoing resistance to Israeli occupation—as their source of pride, while 41% grounded their pride in Palestine's cultural heritage and religious significance. However, across both categories, political dysfunction within the Palestinian Authority dampened national sentiment. Of those who expressed low levels of pride (12.5%), nearly all cited political causes—especially internal corruption, disunity, and poor leadership.

Conclusion

This chapter mapped the expressions of national attachment among Ukrainian, Syrian, and Palestinian respondents, highlighting key patterns within each case. In terms of national self-identification, the majority of respondents across all three samples reported identifying primarily with their national in-group. The Ukrainian sample stands out with near-unanimous identification with the Ukrainian national identity (94%), followed by the Palestinian sample (72%). In the Syrian sample, a smaller majority (53%) identified primarily with their national identity—a finding consistent with existing studies on Syria that highlight the continued salience of ethnic and sectarian identities (Hinnebusch 2020; Phillips 2015). Despite differences in degree, the overall trend across all three cases is that respondents tend to prioritize national over substate or supranational identities.

In terms of levels of national pride, Ukrainian respondents reported the highest levels, with 87% rating their patriotic pride as 'very high' or 'high'. This is followed closely by Palestinian respondents (82.5%), while 67% of the Syrian sample selected the same categories. Conversely, only 2% of Ukrainian respondents rated their national pride as 'low' or 'moderate', compared to 24% of Syrian respondents and 9% of Palestinian respondents. These responses reflect patterns consistent with theories that link national unity to external threats, particularly when the in-group perceives itself as suffering at the hands of a clearly defined out-group. In this context, national groups experiencing ongoing conflict or war with an external actor—such as Ukraine and Palestine—appear to exhibit stronger levels of national cohesion. By contrast, internal conflict, as in Syria's civil war, appears to produce more fragmented expressions of national pride.

When asked about the sources of their national pride, Ukrainian respondents most frequently cited the war effort (71.3%) as their primary reason, while 31% emphasized cultural heritage. These responses suggest that the current conflict with Russia has become a powerful focal point for national unity and pride. In the Syrian sample, cultural reasons were cited most frequently (40%) as the primary source of national pride, with only 18.5% referencing the war effort. This suggests that in the face of civil war, Syrian respondents tend to anchor their patriotism in shared cultural traditions rather than in the ongoing conflict. Palestinian respondents, meanwhile, most frequently identified the 'Palestinian struggle and resistance' (44%) as the basis of their national pride, with cultural reasons coming in a close second (41%). Here, the long-standing conflict with Israel appears to be a central component of national identity and pride, while cultural heritage remains a vital supplementary source.

Although the war effort was a dominant source of pride in the Ukrainian and Palestinian responses, opposition to the war emerged as a significant factor diminishing national pride in all three samples. Respondents who identified as 'not proud' often attributed their sentiments to political dissatisfaction or perceived mismanagement of the conflict. This was most prevalent among Syrian respondents (34%), followed by Palestinian respondents (12.5%), and a smaller segment of Ukrainian respondents (5%). Among the latter, this disavowal of patriotic sentiment was typically tied to political opposition to the war itself or to critiques of the Ukrainian political elite. These findings illustrate that war can simultaneously strengthen and undermine national pride, depending on individual political views and the nature of the conflict. In the Syrian case—marked by civil war, unclear war aims, and fractured factions—respondents were more likely to report decreased pride. In Ukraine, where the conflict is framed as a defence against a foreign aggressor, national pride has intensified. In the Palestinian case, while resistance against an external adversary has fostered cohesion, internal critiques and frustrations have also tempered national pride.

These patterns align with theoretical predictions about the relationship between nationalism and war. The type of conflict experienced appears to shape national sentiment. Wars that pit an out-group against a unified in-group—particularly bilateral conflicts or cases of external aggression—tend to increase group cohesion (see 'conflict-cohesion theory', Levy and Thompson 2010, 100–101). This is especially evident in the Ukrainian case. The perception of an existential threat from Russia has catalysed a strong sense of nationalism, with many Ukrainians reporting that their patriotism was forged or intensified by the shared war effort. In this sense, Ukraine exemplifies the classic model in which war constructs national unity by solidifying in-group boundaries (Gibler et al. 2012; MacMillan 2021).

A similar dynamic is present in the Palestinian case. Decades of protracted conflict with Israel have served to reinforce national identity and a sense of collective struggle. Respondents who perceive the conflict as an ongoing assault on their in-group often derive their pride from the resilience and persistence of the Palestinian people in resisting occupation.

The Syrian case presents a different picture. Prolonged civil war, marked by internal fragmentation and contested political authority, appears to have eroded national pride for many. Yet, cultural pride remains a durable point of identification. Syrian respondents frequently emphasized attachment to Syria's rich historical legacy and shared traditions—rather than allegiance to the state or current regime—as the basis for their patriotism. Even among those expressing political disillusionment, a belief in Syria as a nation endures, grounded in an inclusive historical narrative that transcends religious or ethnic divisions and stands apart from the regime's top-down national discourse. Taken together, the Syrian responses reflect a form of patriotism that persists despite civil war and fragmentation, rooted less in contemporary politics than in a shared historical imagination.

Across all three cases, these evolving patterns of national attachment under conditions of war provide the symbolic and emotional foundations from which people draw when they articulate their visions of the future (Chapter 5).

5
Visions of the Future

Chapter 5 investigates the ideas of the future held by the Ukrainian, Syrian, and Palestinian sample groups. It explores people's visions of their country's future by collecting their views on their home country, its borders, political system, and religious and ethnic landscape, and their personal wish to live in or return to their countries post-war.

When respondents talk about their national communities in the future, they present their idealized versions of the collective to which they belong (Bachleitner 2022; Berenskoetter 2014; Chaitin and Avlagon 2025). While their visions contain different political, ethnic, and religious ascriptions to the nation and its people, ideas of the national community can broadly be distinguished into 'civic' or 'ethnic/religious'. These distinctions stem from the differentiation between 'civic' and 'ethnic/religious' nations (Reeskens and Hooghe 2010). In the literature on nationalism, a civic logic underwrites what Meinecke's (1970 [1907]) typology of nationalisms calls a '*Staatsnation*', a nation which bases belonging on a social construct among members of the 'society' (*Gesellschaft*). In contrast, an ethnic logic underwrites a '*Kulturnation*', or a nation which is founded on a shared heritage among the members of a 'community' (*Gemeinschaft*) (Tönnies 1955). In a *Staatsnation*, membership is outlined in civic terms, whereas in a *Kulturnation*, it is defined in ethnic terms.

Thus, a 'civic' logic of the future nation assigns group membership based on the abstract concept of citizenship. This rationale locates national belonging in shared ideas and political creeds. It defines the boundaries of the community broadly and inclusively, often by invoking criteria rooted in shared humanity, as well as principles of liberty and equality derived from legal norms and the formal definition of citizenship. In contrast, an 'ethnic' logic of national membership confers belonging in more particularistic terms—based on pre-existing characteristics, religious beliefs, and traditions rooted in ancestry and descent (Calhoun 1992 and 2007; Eisenstadt and Giesen 1995, 80–82; Risse-Kappen 2010, 28 and 36; Valadez 2007).

Building on these theoretical distinctions, this chapter presents the sample groups' views on the future, broadly categorizing them according to 'civic'

Legacies of War. Kathrin Bachleitner, Oxford University Press. © Kathrin Bachleitner (2026).
DOI: 10.1093/9780198982913.003.0006

and 'ethnic' logics. Respondents who describe their community in civic terms typically include as members of the national community all those who 'live on a certain territory', 'have legal citizenship status', 'express the will to join the political community', or 'adhere to the basic state ideology'. In contrast, individuals whose responses follow an ethnic rationale are more likely to emphasize 'ancestry and descent', often from the perspective of the 'dominant ethnic/racial/sectarian group' (Shulman 2002, 559).[1] These logics may also underpin different political preferences: a civic logic may translate into support for democracy and federalism, whereas an ethnic or religious logic may manifest in a desire for a strong, centralized state ruled by the ethnic majority.

While 'civic' and 'ethnic/religious' visions of community represent broad archetypes, these categories still offer valuable insight into the kinds of futures the sample groups imagine for their respective home countries. Moreover, by allowing participants to express their visions of the future in their own words, this chapter accounts for the specific contexts and nuances that shape what a civic or ethnic/religious vision means in each case.

Ukrainian Visions of the Future

The survey asked Ukrainian respondents about their ideal future, including preferences regarding Ukraine's borders, political system, regional integration, and ethnic and religious make-up. Beginning with border preferences, a large majority (85%) stated that a future Ukraine must exist within the borders established in 1991. This position implies that Russia would need to return all territory it has occupied since 2014. A small minority (4.5%) supported expanding Ukraine's borders beyond those of 1991, typically referring to Ukrainian-speaking areas within Russian territory. An even smaller group (2%) favoured open borders, while only four respondents (1.4%) said they would accept any borders determined by the outcome of the war (see Figure 5.1).

When asked about their preferred political system for Ukraine's future, the vast majority of respondents (90.7%) favoured a *democratic* system. In contrast, a small minority (7.9%) preferred a *centralized* state, while only 1.4% supported a *socialist* system. These responses suggest a strong collective commitment to democratic governance among Ukrainian respondents despite the ongoing challenges of war and political instability (see Figure 5.2).

Among those who favoured a democratic future for Ukraine, most expressed support for a parliamentary-presidential republic, often coupled

[1] Please note that the terms '*civic*' and '*ethnic*' here refer to the definitions given, and do not imply the (often loaded) connotations and meanings they carry in other contexts.

As of 1991	244	
Expand beyond 1991	13	
As per the outcome of the war	4	
Open borders	6	
I don't know	20	
Total	**287**	

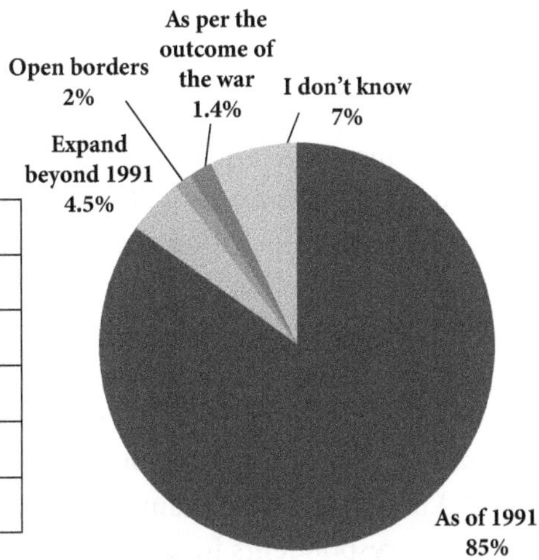

Figure 5.1 Ukrainian Respondents' Preferences for Future National Borders

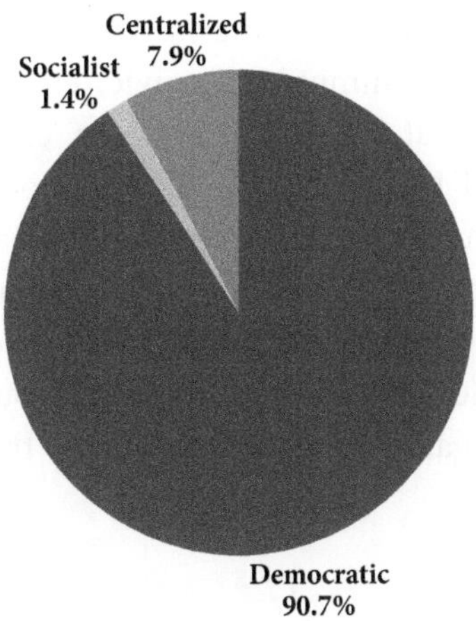

Democratic	253	
Socialist	4	**N = 279**
Centralized	22	

Figure 5.2 Preferred Political System for Ukraine's Future

with the desire for a federal system and EU membership. Their responses frequently emphasized inclusivity, regional autonomy, and anti-corruption reforms. For instance, a twenty-five-year-old woman from Crimea, now living as a refugee in Austria, stated: '*Ukraine should have a parliamentary government with strong regional bodies. I would like to see Ukraine as a country that accepts and facilitates the lives of people of different origins*.' Similarly, a twenty-nine-year-old woman from Kharkiv, currently a refugee in Estonia, noted: '*We must remain the same democratic country with freedom of speech, but we must push out corruption and solve some social and economic issues*.'

In contrast, the 7.9% of respondents who supported a centralized state expressed a preference for *'a unitary, independent state'*. Such a state, as an eighteen-year-old man from Kramatorsk, currently living in Vinnytsia, explained, means *'a centrist regime for Ukraine, one which satisfies all groups of the population. However, without radicalism'.* Only four respondents overall expressed a preference for a socialist system. Among them was a fifty-three-year-old man from Kiev, who described his ideal government as a *'Ukrainian national constitutional labour monarchy and a unitary monostate'.*

Following their political system preferences, respondents were asked about Ukraine's potential membership in political unions. A large majority (89.5%) hoped that Ukraine would join the EU (see Figure 5.3). A smaller share (10.5%) preferred that Ukraine 'not join any union', and notably, none of the Ukrainian respondents in this sample supported joining a customs or military union with Russia or other Commonwealth of Independent States (CIS) countries.

Finally, respondents' visions of Ukraine's future were analysed in terms of whether they conceptualized the national community as civic, ethnic, or religious in nature. Following the theoretical distinction between 'civic' and 'ethnic/religious' nations (Reeskens and Hooghe 2010), a civic model bases group membership on the abstract principle of citizenship, whereas an ethnic logic defines belonging through primordial characteristics, religious affiliation, and inherited traditions. In the Ukrainian sample, a majority of respondents (64.2%) framed their understanding of the Ukrainian in-group in political terms, including references to legal citizenship—responses that

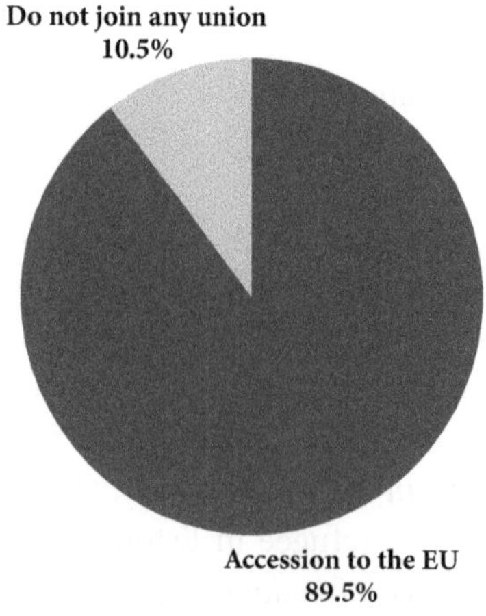

Accession to the EU	257
Do not join eny union	30
Total	**287**

Figure 5.3 Preferred Direction of Political Integration Among Ukrainian Respondents

were coded as 'civic' in the qualitative analysis (see Figure 5.4). The remaining 35.8% of responses were assigned the 'ethnic/religious' code, reflecting views of Ukrainian identity grounded in either 'ethnic' (2.4%) or 'Christian (Orthodox)' (33.4%) orientations.

Responses that reflect a 'civic' understanding of what it means to be Ukrainian are generally more inclusive in their notions of belonging. A sixty-three-year-old woman from Kiev responded that the Ukrainian in-group is for everyone '*who considers him/herself Ukrainian*'. Another woman, a thirty-year-old from Novokrasnivka in the Mariupol region who currently lives in the Cherkasy region, stated: '*To be Ukrainian is not about ethnicity, it's about the state of mind*'. In these responses, which were assigned the civic code, national belonging is tied to certain political ideas: '*In the Ukrainian political nation, ethnicity doesn't matter*', stated a thirty-six-year-old man from Kyiv who currently lives in Perth, Australia. Similarly, a twenty-two-year-old man from Lviv said: '*Ukrainians are diverse, and as a community, it includes all those who express respect and love for Ukraine, its language, and culture. Most importantly, they must identify themselves politically as Ukrainians*'.

These more detailed responses help clarify what it means to '*politically identify as Ukrainian*'. Many in this category expressed a strong commitment to democratic governance (with most favouring a parliamentary-presidential system), support for EU membership, and a desire for a secular state—or a predominantly Christian/Orthodox state that upholds freedom of religion—capable of peacefully integrating all ethnic groups living in Ukraine. As one

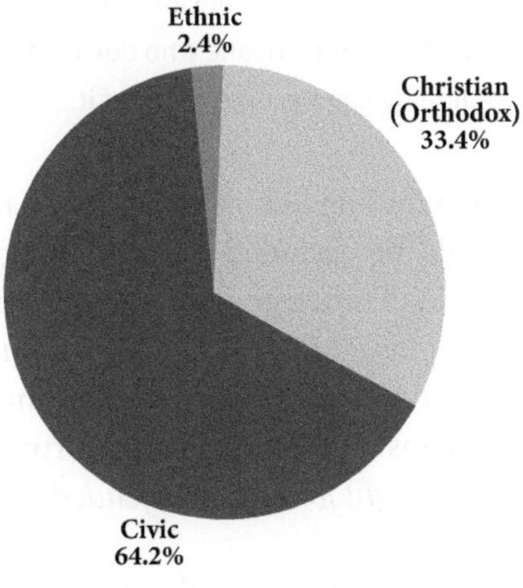

Christian (Orthodox)	96	N = 287
Ethnic	7	
Civic	184	

Figure 5.4 Respondents' Visions of Ukraine's Future: Civic vs Ethnic and Religious Logics

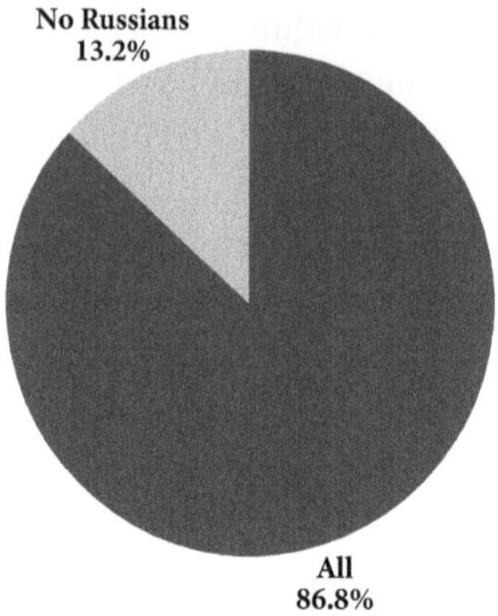

All	249
No Russians	38
Total	**287**

Figure 5.5 Respondents' Perspectives on the Inclusion of Russians in Ukraine's Future

twenty-year-old male respondent from Sumy put it: *'Anyone who shares the pro-European vector'.*

However, while an overwhelming majority of Ukrainian respondents (86.8%) viewed *'all ethnic and religious groups'* as part of Ukraine, 13.2% explicitly advocated for the exclusion of Russians (see code 'All' and 'No Russians' in Figure 5.5). For instance, a thirty-year-old man from Cherkasy, now living as a refugee in Germany, stated:

> The only absolute restriction on residency or even entry into the country should apply to Russians (regardless of their views, political status, or social position) and to Belarusians as long as Belarus is governed by an authoritarian leader who is an ally of Russia. I do not consider Ukrainians as an ethnic group. I consider Ukrainians to be part of a multicultural and multi-ethnic nation, and anyone who does not belong to societies which are openly hostile to Ukraine can become a part of it.

Often, however, respondents who support excluding Russians clarified that only those who do *not* accept Ukraine should be excluded—thus defining Russians in political, rather than ethnic, terms. As a nineteen-year-old woman from Kharkiv put it: *'Everyone who is currently in Ukraine is Ukrainian. However, Russians living in Ukraine should not support Russia's policy'.* Similarly, a twenty-three-year-old man from Berdiansk, currently living in Irpin, stated: *'Without Russians, except for those who fought for us and defended our interests'.*

These responses further illustrate how political affiliation, rather than ethnicity or religion, often defines what it means to be Ukrainian. A similar logic frequently applied to religion. As a thirty-year-old man from Cherkasy, now living as a refugee in Germany, explained:

> I believe that religion plays an important role in the life of Ukrainian society, which is still partially patriarchal and conservative in ethical matters. However, Russia actively exploits this situation. Representatives of the Moscow Orthodox Church are destabilizing Ukraine, and thus they should be banned without exception. Apart from banning the Moscow Orthodox Church, I believe that there should be complete freedom of religion in Ukraine.

Similarly, a twenty-year-old female respondent from Kiev advocates for '*a secular state, BUT all remnants of the Ukrainian Orthodox Church–Moscow Patriarchate must be eradicated*'. She supports a multi-ethnic Ukraine that welcomes '*any ethnic groups*', with one exception: '*Russians who do not pass the citizenship test should be prohibited from coming to us*'. In her view, exclusion is not based on ethnicity but on political loyalty and national commitment.

Some Ukrainian respondents explicitly argued that excluding Russians is not inherently 'undemocratic'. As one eighteen-year-old woman from Kharkiv, now living as a refugee in Rheinsfelden, Germany, stated:

> I want Ukraine to join the European Union, to be a democracy and conduct 'Ukrainization', which means banning pro-Russian parties. Such a step is not contrary to democracy because democracy does not function effectively when it comes to extremism. The ethnic-religious composition of society is of no fundamental importance; rather a future Ukraine should be one without the Moscow Patriarchate and located within the borders approved in 1991.

These responses exemplifies how the Ukrainian in-group is not primarily defined by its confessional identity but by its political orientation. Accordingly, most respondents expressed support for tolerance towards all religious denominations—with the notable and repeated exception of the Moscow Patriarchate.

Overall, while a majority of respondents (64%) described Ukraine's future—and the boundaries of the Ukrainian in-group—in political terms (coded as 'civic'), 36% emphasized specific ethnic or religious attributes as central to defining national identity. Most of these respondents (33.4%) identified 'Christianity', particularly in its Orthodox form, as a key marker of belonging (see Figure 5.4).[2] Yet even within these responses, *ethnic/religious* and *civic* logics often overlapped. Many respondents combined a religious or cultural definition of Ukrainian-ness with a democratic, pro-Western, and European vision of Ukraine as a multi-ethnic state.

Only a small minority (2.4%) articulated a vision of Ukraine rooted in strictly ethnic—and ethnically exclusive—terms. Among these, the label '*Galician*' appeared twice. One respondent, a thirty-five-year-old woman from Kharkiv, called for '*a Slavic ethnos*'. Another, a twenty-five-year-old man from the Sumy region, stated that the nation should consist of '*Ukrainians and loyal minorities, without non-white migration*'. Three respondents explicitly requested a '*mono-ethnic state*', including a forty-nine-year-old man from Irpin who defined that state as one that should include only '*Hutsuls*' and '*Polishchuks*'.

After describing their visions for Ukraine's future, respondents were also asked where they wanted to live—and, for those displaced by war, whether they intended to return. An overwhelming majority (86%) stated that they wished to live in or return to Ukraine, while only 6.3% rejected this option, and 7.7% remained undecided (see Figure 5.6). These responses reflect not only a strong attachment to the idea of a sovereign and democratic Ukraine

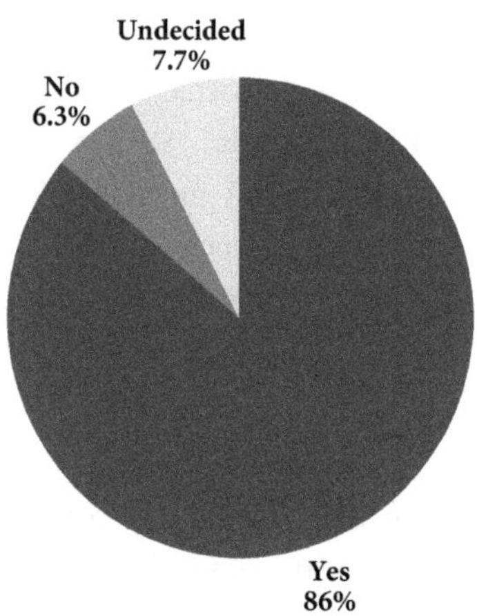

Yes	246
No	18
Undecided	22
Total	**286**

Figure 5.6 Respondents' Intentions to Return to or Remain in Ukraine

[2] In the analysis, responses referencing '*Christianity*' and the '*Orthodox Church*' were grouped together. This is because, in the Ukrainian context, 'Orthodox' typically signifies a preference for Christianity as a religion, with Orthodoxy identified as the respondent's specific denomination. Since the Orthodox Church is one of several Christian denominations present in Ukraine—including Greek Orthodox and Catholic—references to 'Orthodox' were understood as expressions of Christian identity, rather than as distinct from it.

but also a collective desire to rebuild and belong, even amid displacement and uncertainty.

Syrian Visions of the Future

In the Syrian sample, respondents shared their ideal vision for Syria's future. When asked about future borders, a majority (55%) preferred to maintain the nation's current boundaries, established under the 1950 constitution (see code 'national' in Figure 5.7). This suggests that, despite the ongoing civil war, most respondents do not believe Syria's borders should be altered. As a sixty-seven-year-old man from the northern Hama countryside, now living in Australia, explained: '*Syria must preserve its borders as a unified state as they were before the revolution in 2011*'.

Responses in this category often emphasized the '*preservation of Syrian unity*', as stated by a sixty-one-year-old man from Al-Hasakeh, currently internally displaced in Qamishli. Similarly, a fifty-eight-year-old man from Daraa, now a refugee in Jordan, wished for '*the entire territory of Syria without division or annexation of other countries, whoever they are*'. A seventy-five-year-old man from Afrin, currently living in Germany, echoed this sentiment more explicitly: '*The borders are to remain as they were after independence. That means that Turkey, Iran, and all foreign powers must leave Syria, including Russia*'.

While respondents in the *national* category did not want Syria's borders to change as a result of the ongoing civil war, 23.8% of the Syrian sample expressed a 'political' vision of Syria's territorial future (see code 'political' in

National	79
Political	34
Federal	10
Open	8
Separation	8
Arab	4
Total	**143**

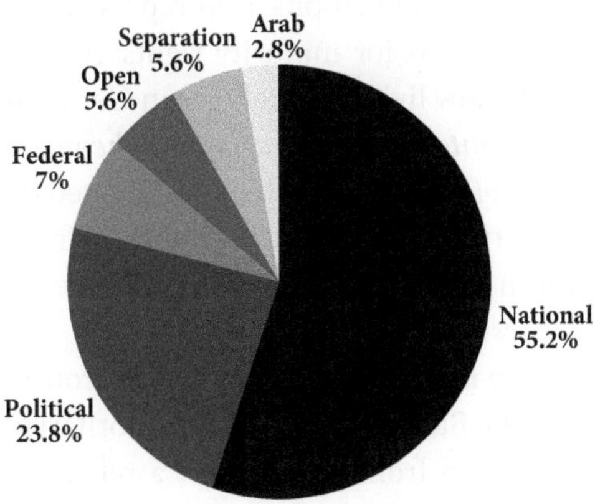

Figure 5.7 Syrian Respondents' Preferences for Future National Borders

Figure 5.7). These respondents supported reclaiming occupied territories—specifically the Golan Heights from Israel and the Iskenderun (Alexandretta) from Turkey. As a thirty-two-year-old man from Damascus, now living as a refugee in Germany, explained: '*Syria must exist within the republic's borders, alongside the restoration of lands colonized by Israel and the pursuit of Arab unity*.' Many of these respondents framed their visions in pan-Arab terms, rejecting the legacy of colonial boundary-making. In the words of a sixty-four-year-old woman from Qamishli, now residing in Australia: '*Syria's natural borders are those before the conflict and before Sykes–Picot*.'

In contrast, 7% of respondents—predominantly those identifying as Kurdish—advocated for Syria to remain within its current borders, but as a unified federal state (see code 'federal' in Figure 5.7). These respondents emphasized the importance of regional self-governance within a broader national framework. As one forty-year-old man from Qamishli stated: '*New regions and administrative divisions are needed which reflect the current circumstances*.'

A smaller proportion of the sample expressed more radical alternatives. Less than 5.6% supported full separation between Syria and its Kurdish regions, envisioning an independent Kurdish state (see code 'separation' in Figure 5.7). Another 5.6% favoured open borders with Syria's Arab neighbours (see code 'open'), while 2.8% advocated for Pan-Arabism—a united Arab nation or, at minimum, '*a true coalition with the Arab countries*' that would span the Levant and reach the borders of Iran, according to a forty-two-year-old man from Damascus currently living as a refugee in Dubai (see code 'Arab'in Figure 5.7).

Regarding Syria's future political system, a majority of respondents (63.4%) envisioned a democratic model (see code 'democratic' in Figure 5.8). These respondents often emphasized representative governance, regular elections, and protections for minority rights. As a fifty-nine-year-old man from Al-Hasakeh, now living as a refugee in Germany, put it: '*I want future Syria to be a parliamentary democracy, where the majority decides, and which preserves the rights of minorities*.' Others highlighted the need for limited political terms, such as a twenty-eight-year-old woman from Damascus, currently residing in Northern Ireland, who stated: '*Elections every 4 years, meaning that a political office does not continue for life*.'

In these responses, democracy was consistently contrasted with the Assad regime, the Ba'ath Party, and its authoritarian political structure. A fifty-eight-year-old man from Daraa, now a refugee in Jordan, envisioned a political system '*which is parliamentarian and has an independent judiciary, an army that does not interfere with politics, and a security apparatus which preserves*

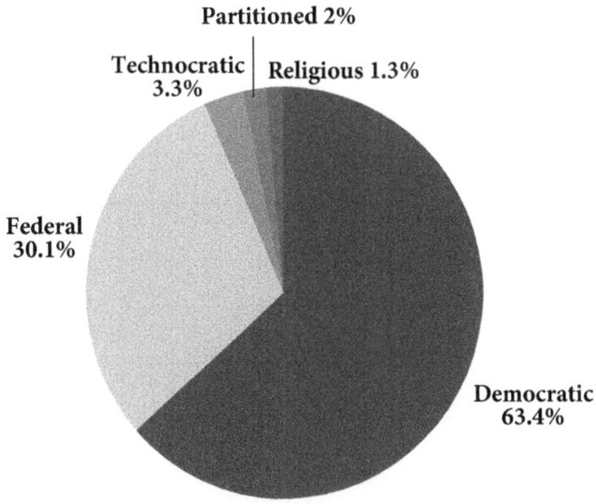

Democratic	97	N = 153
Federal	46	
Technocratic	5	
Partitioned	3	
Religious	2	

Figure 5.8 Preferred Political System for Syria's Future

the security of both the citizens and the state and does not intervene with individual freedoms'.

Furthermore, more than 30% of Syrian respondents stated that the country's future political system should be 'federal', with independent regions. As one thirty-year-old man from Kubani, currently an internally displaced person in Qamishli, explained: '*The Syrian government structure will follow a model of decentralized federal authority, in which all components enjoy cultural, administrative, and political rights*'. Most of these responses came from Kurdish participants. Within this group, 2% went further, advocating for partition (code 'partitioned' in Figure 5.8) to establish an independent Kurdish state.

While many of those who supported democratic or federal models also emphasized the need for separation between religion and politics, a small minority (1.3%) stated that Syria should be ruled by Islamic law (code 'religious'). Additionally, 3.3% of respondents expressed a preference for a technocratic system—an independent government with no political or religious affiliations, committed to rights-based governance (code 'technocratic' in Figure 5.8). As a thirty-three-year-old woman from Damascus, now living in the United Kingdom, put it: '*An independent government that does not belong to any religious or political affiliation*. A thirty-five-year-old man from Homs, now a refugee in Lebanon, envisioned '*a government that guarantees freedom of belief and coexistence while ensuring a fair distribution of wealth and resources*'.

Regarding Syria's future societal make-up, a large majority of respondents (92.2%) envisioned a pluralistic country—a mosaic of ethnic and religious communities living with equal rights (see code 'pluralistic' in Figure 5.9). For

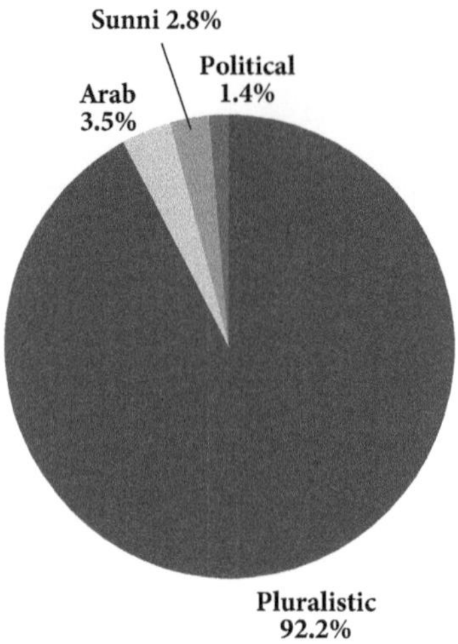

Pluralistic	130	N = 141
Arab	5	
Sunni	4	
Political	2	

Figure 5.9 Respondents' Visions of Syria's Future Ethnic Composition

these respondents, national belonging should be based on shared citizenship and freedoms. As a fifty-year-old woman from Qamishli, currently living in Belgium, explained, this includes *'freedom of belief, freedom of thought and choice'.* Similarly, a thirty-five-year-old man from Derbasia expressed that *'everyone should be equal in their rights and duties'.*

In contrast to the dominant pluralistic outlook, only a small minority of Syrian respondents expressed an ethnic vision of society. Among these, 3.5% believed Syria should consist exclusively of Arabs, while 2.8% stated that Sunnis should be pre-eminent (see codes 'Arab' and 'Sunni' in Figure 5.9). Even within these responses, however, there was a consistent emphasis on granting *'rights to all other religions'*, as expressed by a forty-year-old man from Daraa, now living as a refugee in Germany. Notably, only two respondents (1.4%) argued that future citizenship should be defined politically—based on support for the Syrian resistance (see code 'political' in Figure 5.9). One such respondent, a forty-nine-year-old man from Daraa currently living in Turkey, stated that Syrian society should include *'all groups that did not spill the blood of Syrians'.*

When asked about the desired religious orientation of the future Syrian state, a majority (74.3%) expressed support for a *'secular'* or *'civic'* model (see code 'secular' in Figure 5.10). In Arabic, these terms are often used interchangeably, and many respondents employed *'civic'* to describe a secular state that separates religion from politics. This preference for the term *'civic'* is likely due to its less controversial connotations: while secularism has, in recent decades, been associated in some circles with values perceived to conflict with Islam, *'civic'* is seen as less threatening and more compatible

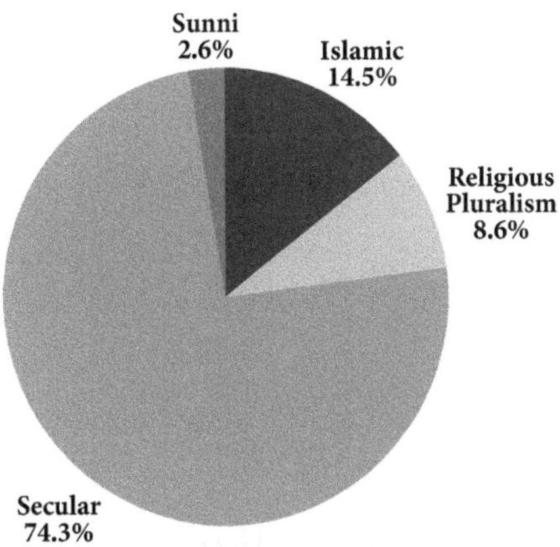

Islamic	22
Religious Pluralism	13
Secular	113
Sunni	4
Total	**152**

Figure 5.10 Respondents' Visions of Syria's Future Religious Orientation

with Syria's historical narrative as a *'cradle of civilizations'.* In contrast, 17% of respondents preferred an Islamic or Sunni Muslim state (see code 'Islamic' in Figure 5.10), while 8.6% supported the idea of a multi-religious Syria (see code *'Religious Pluralism'* in Figure 5.10).

Overall, Syrian respondents' descriptions of their ideal future state and society can be grouped according to a 'civic', 'ethnic', or 'mixed' logic. Visions were coded as 'civic' when they included references to (liberal) democracy; a civic state or society governed by the rule of law and citizenship; separation between religion and politics; a pluralistic and unified society described as a *'Syrian mosaic of sects'* that protects minority rights; and support for free market competition. In contrast, responses were coded as 'ethnic' when they emphasized a Sunni-majority government without democratic institutions; a rule of law without explicit reference to political freedoms; demands for ethnic autonomy or secession; calls to end foreign influences viewed as threats to Syria's unique character; or support for Islamic rule and an Islamic constitution.

As Figure 5.11 shows, a plurality of respondents (48.3%) envisioned Syria's future according to a civic rationale. Another 22.2% expressed an ethnic rationale, while 29.5% of responses reflected a *mixed* logic that combined elements of both.

To illustrate a 'civic' vision of the Syrian state, in their own words, take this response from a forty-nine-year-old man living in Raqqa:

> For my country, I want a constitution that brings all people together. I want the displaced to return to their cities and villages and for us to finally get rid of all the militias and foreigners. I want Syria to have a professional army whose loyalty lies

Civic rationale	85
Ethnic rationale	39
Mixed (Civic & Ethnic rationale)	52
Total	**178**

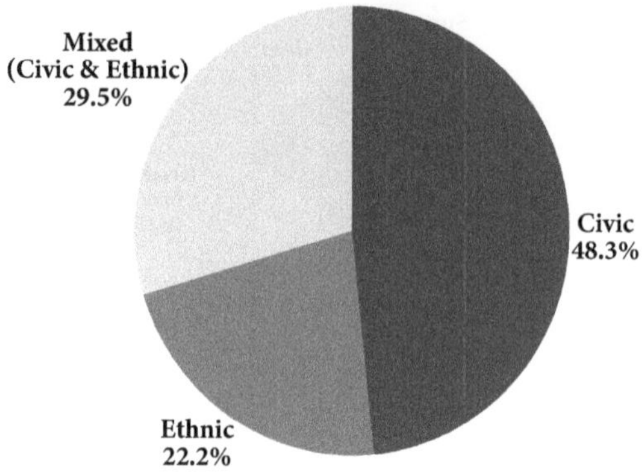

Figure 5.11 Respondents' Visions of Syria's Future: Civic vs Ethnic Logics

with the country, not with individuals. We must restructure some parts of the intelligence services according to the country's needs. Politically, the ideal situation would be a civic and democratic state founded on institutions rather than individuals. I wish for a single and unified state that recognizes all its peoples and the rights of everyone to live under 'the roof of one country'. I want a parliamentarian state with an elected government according to international standards with a message of peace and love instead of a message of hatred and terrorism.

As another example, see the answer from a forty-nine-year-old woman from the coastal region, who shared a similar vision of the Syrian state and society: '*Politically, a secular, democratic regime. Economically, a social market. Religiously, a civic and non-sectarian regime, respecting the rights of the majority and all religious minorities guaranteeing equality to all of them. I want no more confessionalism and not again another military regime!*'

Responses coded under a *civic* rationale frequently included calls to end authoritarianism and military rule in Syria, alongside support for democracy and varying degrees of political freedom. With regard to religion, these respondents consistently advocated for a clear separation between religion and politics, often envisioning Syria as a secular state. As a forty-one-year-old man from Efrin put it: '*Religion is for God, the Homeland is for All! I want a civic, democratic state based on the principle of citizenship and secularism*'.

Conversely, these same respondents often associated sectarianism, factionalism, and Islamism with instability and fear. A forty-year-old woman from Latakia captured this sentiment clearly: '*We need a national project that promotes peace and forgiveness. A national government that encompasses all components of Syria. No more sectarianism, Islamization, regionalism and radical factions!*'

In the Syrian context, the majority of respondents who followed a *civic* logic expressed a clear desire for a democratic Syria. This vision was almost always framed as freedom from military rule and a preference for some form of representative government—if not a liberal, rule-based system led by the people. Many explicitly referred to a '*civic Syria*', which some associated with liberal or civil values such as tolerance, pluralism, individualism, and citizenship, while others linked it directly to secularism. What united nearly all of these responses was a shared desire to preserve Syria's territorial integrity and promote unity among all Syrians. Although this vision was expressed with varying degrees of openness to pluralism, a clear majority described their preferred Syria as '*democratic*' (i.e., representative), '*civic*' (i.e., secular), and '*united*' (i.e., pluralistic).

In stark contrast to this majority view, only a small minority of respondents envisioned an Islamic government in Syria's future. As one forty-nine-year-old man from Deir Al-Zour explained: '*In terms of religion, Syrians are used to simple understandings and applications of religious rules. I prefer such an understanding, which is mainly concerned with abiding by Islamic ethical standards, and which is distant from extremism*'.

Responses such as this—along with others that intertwined religious or ethnic identity with political preferences—were coded as reflecting an *ethnic* rationale. This group, comprising 22% of the sample, varied widely in how they conceptualized the political role of religion or ethnicity. For instance, a thirty-year-old man from Raqqa who identified as a 'Sunni Muslim' stated: '*Politically, I want an arrangement that stops the minorities from controlling the majority. Religiously, however, I want freedom of belief and the freedom for all groups to exercise their religious beliefs*'. Similarly, some Kurdish respondents rejected Arab dominance in ethnic terms. A thirty-five-year-old Kurdish woman from Al-Hasakah said: '*I don't want Pan-Arabism to be part of Syria's future. I am for a federal state in which the Kurds have the autonomy to rule themselves*'.

These views, while reflecting legitimate grievances—particularly among Kurdish respondents historically excluded by an Arab Syrian nationalism—still represent a small portion of the sample. As with those calling for Islamic rule, only a handful within this 22% minority explicitly supported '*ethnic rule*' or '*ethnic separation*'. The overwhelming majority across all categories, including ethnic minorities, ultimately expressed a desire for a '*unified Syria*' within its current state boundaries. As the respondent from Raqqa put it: '*All Syrians belong under the roof of one country*'.

Finally, when asked whether they could envision themselves as part of Syria's future—either by staying or, in the case of refugees, returning—an

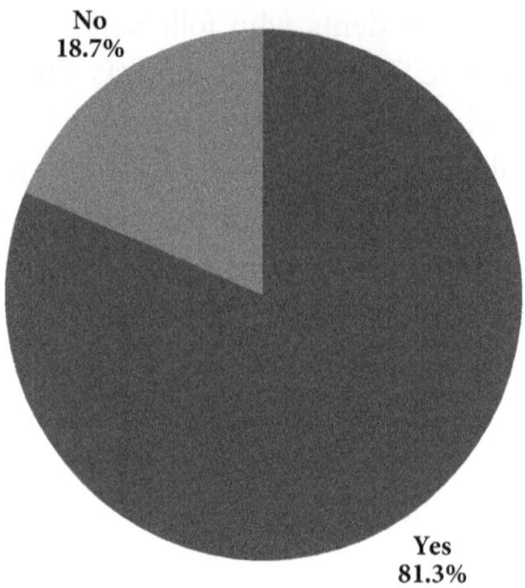

Yes	135
No	31
Total	**166**

Figure 5.12 Respondents' Intentions to Return to or Remain in Syria

overwhelming majority (81%) answered affirmatively. Only 18.7% indicated that they did not wish to return or live in Syria (see Figure 5.12).

Palestinian Visions of the Future

The Palestinian sample offered diverse and often deeply personal visions for the future. Respondents first articulated their preferences for the borders of a future Palestinian state.

A plurality (46%) supported the restoration of Palestine's borders *'from the river to the sea'*, referring to all of historical Palestine prior to the establishment of Israel in 1948 (see code 'Palestine' in Figure 5.13). As a thirty-year-old woman from Gaza declared: *'Palestine is complete from the river to the sea and from Rosh Hanikra Grottoes to Rafah'.* This sentiment was echoed by a twenty-two-year-old woman from Hebron: *'All of Palestine, neither a milli minus nor a plus'.* An eighteen-year-old woman from Ramallah offered a similar vision: *'Palestine from Rosh Hanikra Grottoes in the north, to Rafah in the south, and the Negev, there is no missing inch until the sea: only then, Palestine is complete'.*

By contrast, 30% of respondents preferred a two-state solution within the territory of historic Palestine, envisioning a Palestinian state comprising the West Bank and Gaza Strip (see code 'two-states' in Figure 5.13). For most, this meant a return to the pre-1967 borders at a minimum, though many still expressed aspirations for reclaiming the lands lost in 1948. A forty-one-year-old man from Gaza explained: *'1967 at the very minimum, but do not abandon our occupied lands of 1948'.* A nineteen-year-old woman from Hebron, now

Palestine	40
Two-states	26
One-state	9
Open	8
Arab	4
Total	**143**

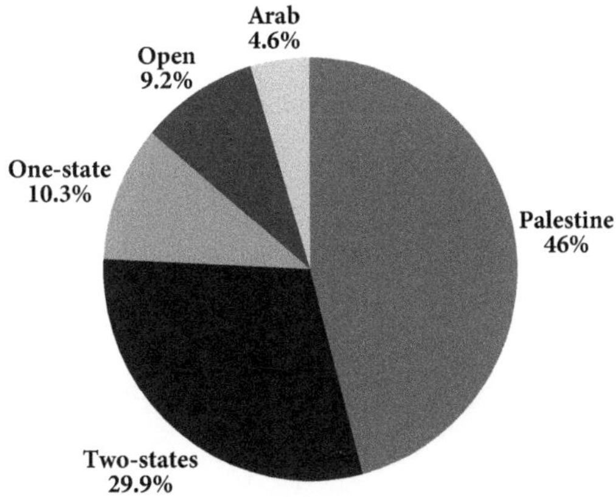

Figure 5.13 Palestinian Respondents' Preferences for Future National Borders

living in the Netherlands, reflected on the tension between hope and realism: '*Ideally '67 borders, realistically (also pessimistically) literally just anything we could possibly get*'. Similarly, a twenty-one-year-old man from the Palestinian diaspora in New York stated: '*1967 borders at the very least. The settlements must be disbanded*'.

Around 10% of Palestinian respondents supported a one-state solution in the form of a binational state, where Israeli and Palestinian citizens would coexist within a shared political framework (see code 'one-state' in Figure 5.13). A slightly smaller group (9%) expressed a preference for *open borders* in a liberal sense—intended to promote freedom of movement, trade, and exchange (see code 'open' in Figure 5.13). Only 4.6% of respondents supported *open borders* within the Arab and Muslim world, in the spirit of Pan-Arabism or Pan-Islamism (see code 'Arab' in Figure 5.13).

Turning to political systems, more than 65% of Palestinian respondents expressed a desire for a future Palestinian state to be democratic (see code 'democratic' in Figure 5.14). Respondents in this category emphasized the importance of periodic elections, the separation of powers, and—most commonly—a presidential system. As a forty-nine-year-old man from Gaza explained, his ideal system would be a '*democratic state that fully preserves the rights of minorities and does not discriminate among its citizens on ethnic or sectarian grounds*'.

In contrast to the majority favouring a democratic system, 18% of Palestinian respondents preferred a strong, centralized, and sovereign state—one that is secular and governed by the rule of law (see code 'sovereign' in Figure 5.14). Another 13.5% expressed a preference for an Islamic state, in which Islamic law would be applied instead of a secular legal framework (see code

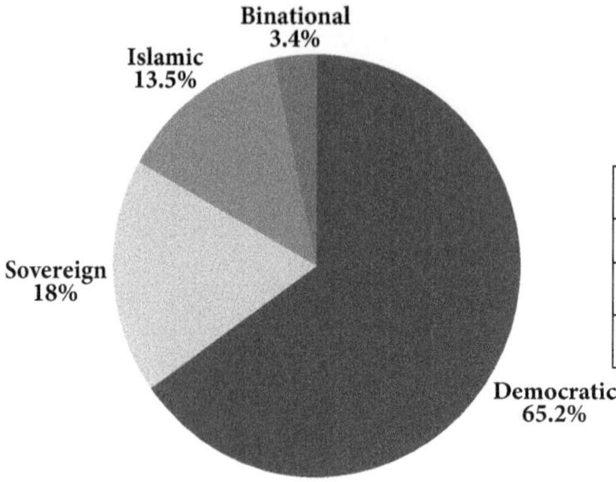

Democratic	58	N = 89
Sovereign	16	
Islamic	12	
Binational	3	

Figure 5.14 Preferred Political System for Palestine's Future

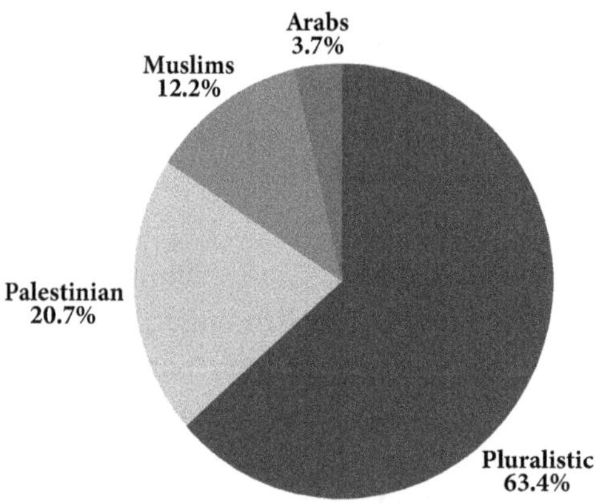

Pluralistic	52	N = 82
Palestinian	17	
Muslims	10	
Arabs	3	

Figure 5.15 Respondents' Visions of Palestine's Future Ethnic Composition

'Islamic' in Figure 5.14). A small minority (3.4%) envisioned a binational arrangement, in which Arabs and Jews would share governance within a single political system (see code 'binational' in Figure 5.14).

Regarding the ethnic composition of Palestinian society, a majority of respondents (63.4%) expressed a preference for a pluralistic Palestine (see code 'pluralistic' in Figure 5.15). What united these responses was the view that the national community should be defined in legal terms—specifically, through the concept of citizenship. These respondents acknowledged the presence of multiple ethnic and religious groups, while promoting values such as diversity, equality, and inclusion. As a fifty-seven-year-old man from Al-Bireh, currently living in New York, explained: *'A state for all its citizens regardless of race, religion, gender, political opinions and sexual orientation'.*

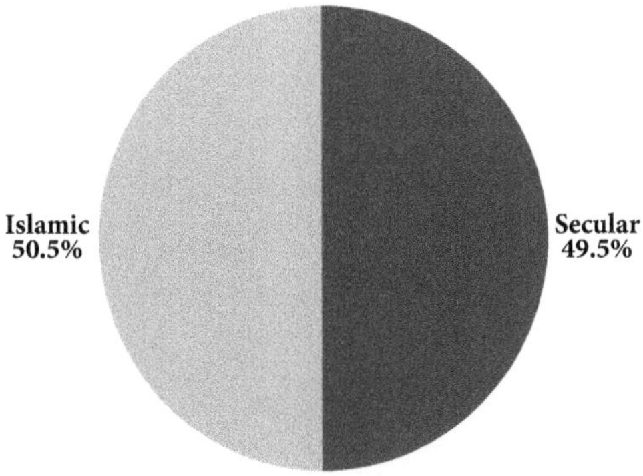

Secular	46
Islamic	47
Total	**93**

Figure 5.16 Respondents' Visions of Palestine's Future Religious Orientation

When asked to define the future ethnic make-up of Palestinian society, 20.7% of respondents referred to the category 'Palestinian', meaning all individuals who inhabited the land prior to the establishment of Israel. According to this view, Palestinian identity is inherited through ancestral connection to the land. As a forty-nine-year-old man from Gaza explained: *'By virtue of the period of time his ancestors lived on this land before the Israeli occupation of Palestine, and he inherited his right to citizenship from father to grandfather'.*

Approximately 12% of respondents defined Palestinians in explicitly religious terms, stating that Palestinians must be 'Muslims' or 'Sunnah'. These responses often rejected ethnic criteria altogether, as expressed by a thirty-two-year-old man from Gaza currently living in Cyprus: *'Islam does not differentiate race'.* A very small minority (3.7%) articulated an ethnic definition of Palestinian identity, envisioning a future state composed exclusively of 'Arabs' (See Figure 5.15).

When asked about their religious preferences for a future Palestinian state, responses were almost evenly split: 49.5% favoured a secular state that safeguards religious plurality, while 50.5% preferred an Islamic state (see Figure 5.16). On this point, the Palestinian sample reveals a near-perfect division between secular and religious visions for the state's future orientation.

As with the other sample groups, Palestinian respondents' visions of a future Palestinian state and society were broadly categorized according to whether they reflected a 'civic' or 'ethnic' logic. Responses were coded as 'civic' when they included explicit references to (liberal) democracy, a civic state or society grounded in the rule of law and citizenship, support for the separation of religion and politics, or the promotion of a pluralistic society in which minority rights are protected. In contrast, responses were coded as 'ethnic' when they

Civic rationale	56
Ethnic rationale	38
Total	**94**

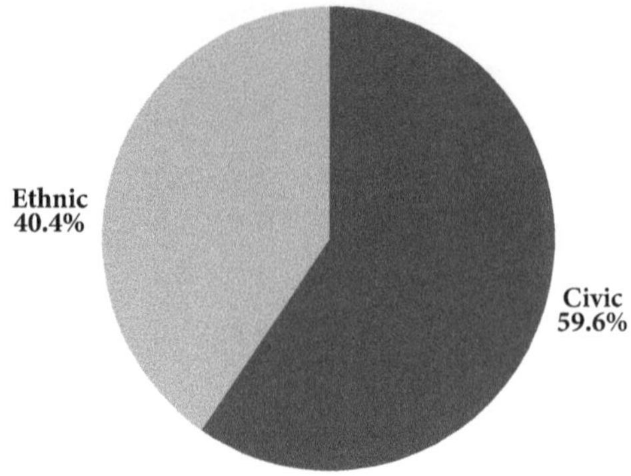

Figure 5.17 Respondents' Visions of Palestine's Future: Civic vs Ethnic Logics

expressed a desire for a homogeneous Arab or Muslim society that excluded others, referenced the rule of law without mentioning political freedoms, or called for the implementation of Islamic rule and an Islamic constitution.

In the Palestinian sample, 59.6% of responses reflected a *civic* vision of the future state, while 40.4% conveyed an *ethnic* vision (see Figure 5.17). For example, a forty-year-old man from Beit Jala in the West Bank described his ideal Palestine as comprising '*at least the West Bank, complete with East Jerusalem and the Gaza Strip*', governed by a '*democratic republican system that safeguards absolute equality between the followers of all religions, and all components of the Palestinian state*'. Similarly, a twenty-one-year-old woman from Ramallah expressed support for '*a Palestinian state in the '67 borders but preferably '48 borders*' with a '*standardized democratic system*', adding that '*religion should not be part of the politics of the state*'. A forty-one-year-old man from Gaza envisioned a '*secular, civic, republican state for Palestine*', and added: '*I think the best at present is the binational state, Arabs and Jews*'.

All of the 59.6% of responses categorized under *civic* logic shared a vision of a pluralistic and democratic Palestinian state in which national identity is defined by *citizenship* rather than *ethnicity* or *religion*. While these respondents commonly advocated for a secular and inclusive political system, their visions diverged with respect to territorial borders. Many adopted a pragmatic stance, supporting a Palestinian state limited to the West Bank and Gaza Strip. However, a significant number also expressed aspirations for a state encompassing all of historic Palestine.

By contrast, the 40.4% of responses coded as *ethnic* consistently referenced a homogeneous Arab and Muslim society, or the establishment of an Islamic government in Palestine. In most cases, these respondents favoured borders that encompassed all of historic Palestine. Some took this further, envisioning the removal of borders altogether within a broader Arab or Islamic union. As an eighteen-year-old man living in Al-Arroub refugee camp put it: '*I don't want there to be borders; because we are all Muslims*'. He envisioned a future state governed by a '*rightly guided caliphate system*' encompassing all '*Sunnah of the Prophet*'.

Other examples illustrate the diversity within the ethnic logic. A twenty-one-year-old woman from Jerusalem, now living in Hebron, imagined a '*fully sovereign Palestinian state*' that '*comprises the borders of historic Palestine*', '*allows for all religions*', but is '*primarily Islamic yet includes all different ethnic groups*'. In her view, ethnic pluralism is compatible with Islam's universal message. A twenty-two-year-old man from Gaza envisioned an '*original Palestine*' that '*includes Muslims and Christians, but no Jews*'. A forty-year-old man from Gaza imagined a '*Palestinian, Arab, Muslim state that embraces all religions*' but remains predominantly Islamic. He defined its borders as '*historical Palestine from sea to the river with all its historical components*', and said it should be open to '*Palestinians and everyone who wants to live under the Palestinian flag*'. Similarly, a thirty-eight-year-old woman from Gaza described a state with '*historical borders, without partitions and barriers*', open to '*Muslims and Christians only*'. She added that it should have an '*independent political system with no competing parties and no political discrimination*', and should be inclusive of '*anyone who is an original citizen in the historical land of Palestine*'.

Finally, Palestinian respondents were asked whether they hoped to live in a future Palestinian state—or, in the case of refugees and diaspora members, whether they wished to return. An overwhelming majority (90.5%) answered affirmatively, while only 9.5% indicated that they did not wish to live in a future Palestine (see Figure 5.18).

Conclusion

Across all three samples, the majority of respondents presented their visions of the future through a *civic* rather than *ethnic* lens. This means that most Ukrainian, Syrian, and Palestinian participants expressed a preference for democratic governance and a national community grounded in citizenship and shared political principles, rather than in ethnic or religious identity.

In the Ukrainian sample, support for democracy was overwhelming, with 90% expressing a preference for democratic governance. Similarly, 64% described their ideal national community as one based on legal citizenship and open to all who adhere to a shared state ideology—defined, in this

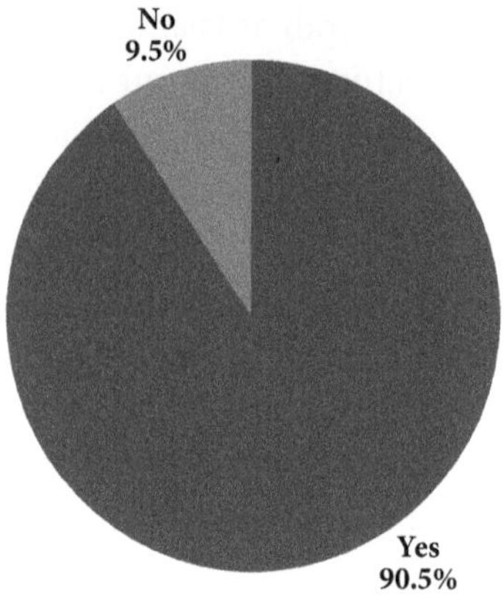

Yes	86
No	9
Total	**95**

Figure 5.18 Respondents' Intentions to Return to or Remain in a Future Palestinian State

context, as democratic and pro-European. While 87% believed that Ukraine's future should include people of all ethnic backgrounds, 13% supported the exclusion of Russians. Notably, however, this exclusion was framed in political rather than ethnic terms. Regarding borders, 85% affirmed the national integrity of Ukraine by supporting the 1991 post-independence boundaries. A further 86% expressed the desire to live permanently in Ukraine. Taken together, these responses define the Ukrainian nation, its people, and its future in explicit opposition to Russia, with national unity framed as the highest social good.

In the Syrian sample, 64% of respondents also supported democracy, and 48% described their ideal society as pluralistic and citizenship-based, with equal rights for all religious and ethnic communities. Another 22% preferred an Islamic state. Regardless of preference, all respondents defined their ideal Syria in opposition to the Assad regime. While many saw the current war as an opportunity for regime change, a majority (55%) insisted that Syria's territorial borders remain unchanged. Additionally, 81% expressed a desire to live in or return to Syria. Overall, respondents defined Syria's future in contrast to Assad's vision, drawing instead on the country's long-standing historical and social traditions. For most, the minimal condition for the future was liberation from Assad's regime; for others, a maximalist vision included a democratic Syria that respects its sectarian mosaic under the rule of law.

In the Palestinian sample, 60% envisioned a democratic, pluralistic, and secular Palestinian state, where national identity is based on citizenship rather than ethnicity or religion. In contrast, 40% supported an ethnic or religious vision, with many advocating for an Islamic government. In terms of borders,

46% of respondents preferred a future Palestinian state encompassing all of historical Palestine, and a large majority (90%) expressed a desire to live permanently in Palestine. Here too, visions of the future were framed in opposition to the political ideals of the perceived enemy—Israel. For many Palestinian respondents, the protracted conflict with Israel has shaped their sense of what a political community should be and who should belong to it. Their ideal future centres on self-determination for all those with roots in historical Palestine—yet with a clear distinction from Jews. This distinction may help explain why half of the respondents favoured an Islamic state.

Across all three samples, the ideal future is consistently defined in opposition to the warring party and its political and societal vision. This contrast is most pronounced in cases involving external aggression, such as Ukraine, where Russia's invasion has galvanized a strong civic, democratic, and pro-European outlook. Ukrainian respondents overwhelmingly defined the nation in political terms that stand in direct opposition to Putin's Russia. Similarly, Syrian respondents defined a civic vision of Syria in contrast to Assad's authoritarian regime—democratic and pluralistic, and in some cases framed through religion. Among Palestinians, the preferred vision was likewise democratic and pluralistic, often Islamic in orientation, and centered on inclusion of all those rooted in historical Palestine, again defined in opposition to the Israeli political project.

Importantly, across all three cases, religious and ethnic preferences were predominantly framed through a political rather than primordial lens. References to religion or culture often served to mark political distinction from an antagonistic 'other', rather than to assert fixed ethnic or sectarian boundaries. Taken together, these findings suggest that under conditions of war, visions of the future are shaped less by inherited identities than by conflict-driven oppositional frameworks through which political community, belonging, and peace are reimagined.

6 Conclusions

Transforming the Legacy of War into Peace?

The book explored the legacy of war left on people and their societies. Through the eyes of the three sample groups—Ukrainians, Syrians, and Palestinians—the empirical Chapters 3–5 allowed respondents to share their experiences of war by recounting, first, their war narratives; second, their levels of national attachment and patriotic feelings towards their countries; and third, their visions of the future. While the aim of this book was to define and map the legacy of war in this way, these three elements of war legacies also offer hints as to how respondents choose to define 'peace'.

As emphasized in the introduction, while respondents' narratives of war, their national attachments, and their visions of the future are compelling in their own right, they are particularly important for International Relations because they offer insights into how societies transition from war and conflict to peace (Chaitin and Avlagon 2025). The theories discussed in this book suggest that people's interpretations hold the potential to generate new political and societal imaginaries, thus envisioning what a sustainable future peace might look like. 'Peace is an idea', and like all ideas, it depends on the choices we make (Coker 2014, 105). Through recounting the war, processing collective emotions towards a country in conflict, and imagining its future, novel social constructions emerge that help define the normative horizon for peace.

In this final part, the chapter reviews the trends that became apparent within the sample groups. These trends may help signal possible implications for both the prospects and the obstacles to a transition to peace.

While it is beyond the scope of this study to test the links between war legacies and peace empirically, future research is advised to examine these links across a wider range of cases, with larger, representative data sets and perhaps through experimental designs. In the meantime, this chapter presents the trends observed in the sample groups. It places these trends in conversation with findings from the interdisciplinary literature to point out potential pitfalls and opportunities for the termination of conflict. Notably, the book deliberately does not seek to settle the question of what peace is—or worse,

Legacies of War. Kathrin Bachleitner, Oxford University Press. © Kathrin Bachleitner (2026).
DOI: 10.1093/9780198982913.003.0007

what it should be. Instead, it works out the visions of peace presented by the participants of the Ukrainian, Syrian, and Palestinian sample groups.

'Peace as Reconciliation': War Narratives as Obstacles and Opportunities

This section synthesizes the dominant patterns that emerged in respondents' war narratives and considers what these interpretations imply for the prospects of peace. Across the three cases, a first and striking trend concerns how respondents make sense of the war itself. When respondents from the three sample groups were asked about their descriptions of the war—its rationale, and their understandings of victims and perpetrators—the majority across all samples narrated the conflict through a primarily 'political' logic of power struggles and interests.

The Ukrainian narrative of the war with Russia was a rather uniform story that positioned Russia as the aggressor, and the Ukrainians as the defenders of the country's independence and national integrity (92%). Furthermore, the majority of Ukrainian respondents attributed all responsibility to Russia (82%), by which they meant not only the Russian leadership but also its citizens. In contrast, the Ukrainian collective is singled out as the victim of this aggression (76%).

The Syrian account of the civil war was one of dictatorship, oppression, and a national struggle for freedom, while global powers intervened in line with their own interests. A majority (71%) described this struggle as shaped by political agendas and power calculations. However, 29% offered accounts that framed the conflict in sectarian terms, portraying it as a case of Alawi tyranny over a Sunni-majority country. While there was strong agreement within the Syrian sample on who was responsible for the war—namely Assad's government and its foreign supporters (86%)—some respondents depicted these actors in explicitly sectarian terms. There was also variation in how victimhood was assigned: 62% identified all Syrians collectively, whereas 38% singled out specific ethnic or sectarian groups, usually Sunnis or Kurds.

In the accounts of the Palestinian respondents, the conflict with the Israelis was described as a political-territorial contest rather than one rooted in religious differences. A majority of Palestinian responses conveyed a relatively uniform narrative of Israeli occupation and Palestinian victimization. Collective victimhood was notably high in the Palestinian sample, at 82%. However, there was considerable variation in the assignment of responsibility for the conflict: while half placed full responsibility on Israel, the other half distributed blame equally between both sides. A small but significant share (13%) assigned complete responsibility to the Palestinian side. In this regard, the Palestinian responses stood out for their self-critical view of the in-group. This self-critique, however, stemmed from two distinct motivations with different potential implications for peacebuilding: some respondents recognized the Palestinian role in escalating the conflict, while others blamed internal Palestinian divisions for hindering the goal of liberating the whole of Palestine.

According to the theories, distinctions in the designation of victims and perpetrators matter for either perpetuating political violence or ending it, as they can plant the seeds for compromise and the opening of peace negotiations. Scholars have shown that framing opposing groups in polarized, black-and-white, good-versus-evil terms exacerbates violence and reduces the chances for peace (e.g. Bar-Tal 2007; Bar-Tal et al. 2014; Bar-Tal et al. 2013; Bar-Tal and Halperin 2013; Canetti et al. 2015). Such narratives tend to conflate entire collectives—victims or perpetrators—without distinguishing between political leaders, actors, and their motivations. Furthermore, these framings often fail to adopt a self-critical perspective on the in-group's role in the conflict (Branscombe 2004, 330). By offering the comfort of simplified binaries, they allow the in-group to distance itself from its own involvement in the violence, avoiding challenges to deeply held assumptions and instead fostering strategies such as scapegoating and revenge-seeking (Schick 2011, 1854).

The distinctions in the way people construct war narratives thus make a difference for the prospects of peace. For instance, from research findings in Social Psychology, we know that in cases where the collective side of victimhood is heightened (Acosta 2021), and where the in-group is viewed as suffering solely under an unjustified aggressor, the in-group's psychological openness to peace with the out-group is significantly lowered (Bar-Tal et al. 2009; Bar-Tal et al. 2010; Chaitin and Avlagon 2025; Kelman 1987; Maoz and McCauley 2005). The strong notion of collective victimhood evident in the Ukrainian and Palestinian samples hence is likely to pose a potential obstacle to finding a compromise with the out-group in any future peace settlement. Moreover, when it comes to the distribution of responsibility, breaking up the clear distinction between victims and perpetrators, particularly by embracing responsibility also for the in-group's harm-doing is said to reduce inter-group prejudice and violence. Any admission of the in-group's role in the harm-doing, however, is usually costly in conflict scenarios and thus avoided (Branscombe 2004, 330). The low numbers of respondents across samples that assign responsibility to the in-group—5% in the Ukrainian sample and 13% in the Palestinian sample—confirm this logic.

While the notion of collective victimhood often hardens divisions in wars between two opposing groups, in the context of civil war it has the potential to unite the in-group. The high level of collective victimhood attributed to 'all Syrians' (62%) in our Syrian sample can be interpreted as a sign of national cohesion in response to the war. When the suffering in-group is understood to mean the Syrian population as a whole, rather than a specific ethnic or sectarian group (38%), this may be seen as a tentative indication that Syria is unlikely to fragment as a result of the civil war.

The ways in which war narratives are constructed—and victims and perpetrators defined—carry implications for the outcome of peace, though these effects vary depending on the type of war. In inter-group conflicts, such as those between Ukraine and Russia or between Palestinians and Israelis, a more differentiated understanding of collective victimhood and perpetration may increase the chances for eventual rapprochement. In civil wars, as illustrated by the Syrian case, the designation of collective victimhood may foster a sense of shared suffering that unites the national collective.

Consistent with these expectations, our findings suggest limited prospects for peace, particularly in the two cases of inter-group war. In the Ukrainian sample, respondents' war narratives were highly polarized, marked by a pronounced sense of collective victimhood in the face of an external aggressor and by the attribution of blame to both Russian political leaders and Russian citizens as a collective.

By contrast, Palestinian war narratives appeared somewhat less polarized. Although Palestinian respondents likewise expressed a strong sense of collective victimization, roughly half engaged in a degree of self-critical reflection, assigning responsibility to both sides. Importantly, blame was directed primarily at political leadership, while Palestinians were portrayed as victims not only of the conflict itself but also of their own leaders. In many cases, however, this self-critique did not address Palestinian contributions to the conflict as such, but rather conveyed frustration with the failure to achieve Palestinian political objectives. Nevertheless, these more differentiated narratives indicate a limited openness to reconsidering established conflict positions.

With regard to the distinction between 'political' and 'ethnic/sectarian' conflict narratives, existing IR scholarship suggests that compromises leading to peace agreements are more likely when conflicts are framed in formal political terms rather than through emotionally charged symbolic narratives grounded in ethnic, sectarian, or ideological patterns of thinking

(Manekin et al. 2018; Hassner 2009; Kaderan 2017; Smith 2000; Toft 2003). Accordingly, respondents who articulate a formal political interpretation of war—depicting actors and their motivations as guided by competing political interests rather than ideological or religious fervour—are more likely to express openness to negotiated settlements (Fearon 1995; Cederman et.al 2011). Across the three samples, a majority of responses reflected this political logic: approximately 80% in both the Ukrainian and Palestinian cases, and 71% in the Syrian case, a context characterized by stronger ethnic and sectarian cleavages.

At the same time, these findings should not be overstated. As the literature itself acknowledges, political and symbolic values are often deeply intertwined in wartime, particularly in territorial conflicts (Manekin et al. 2018). The Ukrainian case illustrates how territory, although not religiously coded, can acquire near-sacred status when framed as national defence against external aggression. Similar entanglements of political and symbolic meanings are evident in the Palestinian context. Consequently, the prevalence of political war narratives in these samples offers only cautious grounds for optimism: political disagreements in polarized conflict settings frequently become infused with ideological and even religious significance, transforming what may appear analytically as a cost–benefit calculation into a struggle over non-negotiable values.

Nevertheless, in all cases, the interpretation of political violence matters because—as Morgenthau (1970, 8–9) put it—'when people see things in a new light, they may act in a new way'. In the context of war, however, this relationship often militates against peaceful change. Wartime logics demand solidarity through victimization, and identities grounded in collective suffering reinforce sharp boundaries between in-groups and out-groups through shared grievances. Conflict narratives framed in this manner also mirror dominant political discourses and broader contextual pressures (Alexander 2012).

These dominant narrative frames are both shaped by and reinforced through political leadership. In polarized wartime contexts, leaders often amplify victimization and moral condemnation, responding to violence with emphatic denunciations that resonate with popular demands for decisive leadership and moral clarity. This dynamic is evident in the rhetoric of Ukrainian President Volodymyr Zelensky, Syrian President Bashar al-Assad, and Palestinian leaders. For example, in his Independence Day address on 24 August 2022, President Zelensky stated: 'What will bring the end of war? We used to say "peace". Now we say "victory". [...] We would rather live in trenches than live in chains. And we will put our hands up only once, when we raise them to celebrate our victory. As the whole of Ukraine. We do not bargain away our lands and our people. Ukraine means all of Ukraine' (Zelensky 2022, 116–117).

Similarly, President Bashar al-Assad, in his Army Day speech in 2011, declared: 'Syria is used to creating victories and defeating the enemies of the country and nation [...] it knows to [...] add new victories and leave warmongers and blood merchants to taste the bitterness of defeat and disappointment'.[1] Polarized rhetoric also appears in the speeches of Palestinian President Mahmoud Abbas, who proclaimed in a pre-recorded speech to the UN in 2020: 'How long will the Palestinian people continue to languish under the yoke of Israeli occupation [...]. In spite of the difficulties and the unjust blockade [...] we will not bow down, we will not surrender, we will not compromise, and we shall triumph'.[2]

Sustained armed conflict relies on such polarized moral logics, in which denunciation of the adversary becomes both politically functional and affectively rewarding. As Schick (2011, 1852) notes, moralistic denunciation provides immediate gratification and temporarily cleanses the speaker of any proximity to guilt, yet only in the short term. While this framing may be effective in sustaining war efforts, it sits uneasily with the kinds of reflection required for any durable transition away from violence. Peace processes—whether at the level of individuals, communities, or political leadership—thus involve a confrontation with the limits of polarized moral frameworks.

It is precisely the problem of how subjects respond to violence, and with what ethical consequences, that Judith Butler interrogates in *Precarious Life: The Power of Mourning and Violence*. Butler (2004, 16–17) asks those affected by violence to reflect on the kinds of political and moral subjects they become in responding to injury:

> What role will we assume in the historical relay of violence, who will we become in the response, and will we be furthering or impeding violence by virtue of the response that we make? To respond to violence with violence may well seem 'justified', but is it finally a responsible solution? Similarly, moralistic denunciation provides immediate gratification and even has the effect of temporarily cleansing the speaker of all proximity to guilt through the act of self-righteous denunciation

[1] Speech delivered on 1 August 2011. The full text was downloaded in 2019 from President Assad's official website, which at the time hosted English translations of all his speeches. As of writing, the website appears to have been taken down. http://www.presidentassad.net/index.php?option=com_content&view=categories&id=81&Itemid=496 (viewed September 2019).

[2] Pre-recorded speech of Mahoud Abbas, at the general debate of the General Assembly's 75th Session, 2 September 2020. See UN News: https://news.un.org/en/story/2020/09/1073712 (viewed 18 September 2024).

itself. But is this the same as responsibility, understood as taking stock of our world, and participating in its social transformation in such a way that non-violent, cooperative, egalitarian international relations remain the guiding ideal?

Of course, such a reflection is profoundly difficult for individuals and communities that have suffered loss and injustice through war. Yet, however painful, critical questioning within and between groups remains essential to understanding the whys of extreme suffering. All reflection and insight into the practices that facilitate violence and trauma are therefore crucial: through enhanced understanding, it may become possible to challenge the damaging, knee-jerk reactions that perpetuate the cycle of violence (Schick 2011, 1854–1855). The scope for reflection and reinterpretation in the wake of trauma thus leaves a glimmer of hope that, if successful, it can give rise to a form of 'peace' understood as a 'reconciliation of different values'. Only peace conceived in these terms is likely to prove sustainable over time.

'Exhausted Peace': Nationalism and Conflict Transformation

A second, related dimension concerns respondents' levels of national and patriotic attachment and their implications for the potential for peace. When asked about their feelings of national attachment and pride, Ukrainian respondents expressed a strong sense of national identity: 87% described their pride in the nation as very high or high. Among Syrians, national attachment also emerged as a prominent theme. Despite years of civil war and deep internal fractures, 67% reported feeling very high or high levels of national pride, suggesting a continuing emotional connection to the idea of Syria as a national entity. In the Palestinian sample, nationalism and patriotism were equally pronounced. Here, 82% of respondents expressed very high or high levels of pride in the nation, reflecting a strong affective bond in the face of occupation and political uncertainty.

Furthermore, the Ukrainian sample not only exhibited high levels of patriotic pride, but also a significant number of respondents (71.3%) identified the war effort as their primary source of pride. In the Palestinian sample, national pride was similarly grounded in political struggle, with 44% citing the Palestinian resistance as the main reason for their pride. In contrast, the Syrian sample—where overall levels of patriotic pride were lower—showed a different pattern: 40% of respondents based their national pride on cultural reasons, such as history and tradition, while only 18.5% linked it to the war effort. Uniquely within the Syrian group, 34% explicitly named the civil war as the reason why they were not proud to be Syrian—a tendency not present in the Ukrainian sample and only marginally present (12%) in the Palestinian responses.

When applying the conceptual distinction between nationalism and patriotism, we find that high levels of national pride rooted in the war effort can be described as 'nationalistic', where nationalism implies asserting one's nation, often in opposition to others. By contrast, national pride based on culture, tradition, or historical continuity—without a direct political or exclusionary framing—can be described as 'patriotic'. Patriotism, in this sense, reflects love for one's country that does not negate or diminish others (Viroli 1997; Wimmer 2019). Mapping these definitions onto the sample groups, the qualitative analysis of the responses suggests that the Ukrainian and Palestinian samples reflect high levels of nationalism, whereas the Syrian sample reflects higher levels of patriotism.

These trends are not surprising when considering the types of conflict these groups experience, and what the theories predict about the links between nationalism and war. Wars, particularly in the form of bilateral conflict, pitch an out-group against an in-group increasing the cohesiveness of the in-group (see 'conflict-cohesion theory', Levy and Thompson 2010, 100–101). As political scientists explain it, people 'rally around the flag' in the face of outside attacks, mainly so as not to endanger the war efforts of their in-group (Baker and Onea 2001). These tendencies are, without doubt, present in the Ukrainian case: the Russian external aggression sparked a strong Ukrainian nationalism that came to the defence of Ukraine as a nation. An in-depth qualitative analysis of the Ukrainian responses further reveals that in many cases, people only discovered their national pride as a result of the united war effort. The Ukrainian case, in that regard, fits the textbook example of how war creates nations by cementing national in-groups (see, for instance, Gibler et al. 2012; MacMillan 2021). However, similar forces are at work in the Palestinian context of a protracted, generational conflict with the Israelis. Palestinians, over the decades, have become strongly united as a national group in the defence of their homeland. The majority of Palestinian respondents who interpret the conflict as a constant attack on their in-group also base their pride first and foremost on Palestinian persistence and perseverance in the face of occupation.

Notably, the dynamics between nationalism and war differ in the context of civil war, where the in-group is threatened from within, typically along ethnic, sectarian, or political lines. As a result, the expected link between war and nationalism in such cases often points to rupture rather than cohesion. Unlike interstate wars, civil wars tend to fragment national entities into smaller ethnic or sectarian units (Ayoob 2007; Bates 2008; Kaufmann 1996b; Rotberg 2004). This context helps explain why the trends observed in the Syrian sample deviate significantly from those found in the Ukrainian and Palestinian groups. The chaos and grievances that follow in the wake of civil war contribute to the lower levels of national pride among Syrian respondents and their stronger reliance on cultural—rather than conflict-based—sources of national attachment.

In their responses, many Syrians emphasized that a shared historical culture continues to unite them despite, and irrespective of, the civil war. Crucially, this finding challenges the theoretical expectation that the Syrian national community might fragment. Even respondents who reported losing their national attachment due to the country's political trajectory expressed a belief in a future Syria that differs fundamentally from Assad's. They envisioned a Syria that is rooted in ancient historical traditions rather than in Assad's imposed political framework. Such a vision of Syria includes all Syrians, regardless of religious or ethnic identity. The in-depth qualitative analysis of the Syrian sample thus revealed a heightened sense of patriotism—formed because of, or in spite of, the civil war—that contrasts with the nationalism observed in the inter-group conflict scenarios.

In line with interdisciplinary research on war and its effects on national attachment, the trends observed across the three groups suggest that both patriotism and nationalism tend to surge in response to political violence (Wimmer 2014; Wimmer and Min 2009; Bar-Tal et al. 2014). In our samples, whether individuals became more nationalistic or more patriotic appeared to depend on the type of war they experienced. In cases of external aggression, nationalistic sentiments were most pronounced among those who supported the war effort. Within this nationalistic logic, the in-group consolidates in opposition to a perceived evil out-group. In contrast, patriotism prevailed in the Syrian sample. Amid the political chaos triggered by the civil war, respondents expressed attachment to the idea of Syria primarily through cultural reference points—historical continuity, tradition, and identity—rather than current political realities.

The rise of nationalism and patriotism in wartime, however, appears strongly shaped by individuals' political orientations. Among those in political opposition—whether due to ideological disagreement or reluctance to make the personal sacrifices war demands—politics emerged as the primary reason for national disaffection across all sample groups. At the time of data collection, this form of disaffection was lowest in the Ukrainian sample (5%), somewhat higher in the Palestinian sample (12.5%), and significantly more prominent in the Syrian case: 34% of Syrian respondents stated that the war and the country's political situation had caused them to lose their national attachment, and with it, any sense of pride.

What do these trends imply for the prospects of peace? Theories that link wartime surges in nationalism to intensified conflict suggest that heightened nationalism may exacerbate hostilities rather than resolve them. While nationalism may be defined as 'the exclusionary love of one's country at the expense of others', it often becomes especially radicalized in the polarized contexts of political violence. Social psychological research shows that the high levels of sacrifice demanded during war often lead to glorification of the in-group and dehumanization of the out-group (Bar-Tal 2007; Bar-Tal et al. 2013; Vollhardt and Bilewicz 2013). As a result, while war may strengthen internal solidarity, it simultaneously hardens boundaries with the out-group, reinforcing hostility. Nationalism that intensifies as a result of war is therefore deeply conflict-prone and, by its nature, not conducive to peace (Hutchinson 2017, 160–184).

In cases of civil war, the relationship between nationalism and conflict is less clear-cut. Civil wars—such as the one in Syria—tend to rupture the national collective, often resulting in diminished nationalistic tendencies. However, as our findings have shown, a form of patriotism grounded in cultural identity persisted. By rooting their pride in Syria's historical identity as the cradle of civilizations, many respondents expressed patriotic attachment that endured irrespective of the ongoing civil war. For the prospects of peace, this suggests that where strong patriotic sentiment connects individuals to a shared cultural heritage—beyond the divisive politics of the present—such a wave of cultural patriotism, as observed in Syria, may help preserve rather than fracture the national in-group in the aftermath of civil war.

In contrast, the findings from the Ukrainian and Palestinian samples give reasons to assume that these conflicts will persist, especially if they keep being fuelled by high levels of nationalistic sentiments, which are sustained by respondents' pride in the war efforts. On the contrary, this implies that, where political opposition to the war effort rises, either out of growing war weariness or political opposition to the leadership, the levels of nationalistic fervour will diminish. However, for such a 'war weariness' or 'war fatigue' to emerge, it usually takes a long time and an astonishingly high toll on civilian, military, and financial resources (Greig 2008; Hosli and Hoekstra 2013). Such a sentiment was evident in neither the Ukrainian nor Palestinian samples. Despite the exacerbation of both conflicts since the time of the surveys, it also does not seem evident from opinion polls today.[3]

Moreover, placing hope in the emergence of war weariness as a counterforce to the nationalistic sentiments generated by war is neither ethically sound nor likely to produce a sustainable peace (Blainey 1988, 9; Cashman 2013, 232). If anything, such war weariness may result in what can be termed an 'exhausted peace'—a fragile and temporary reprieve rather than a durable resolution. Historical examples illustrate the dangers of this dynamic. Germany in the aftermath of the First World War, for instance, experienced a brief period of subdued nationalism, only for nationalistic fervour to return in an even more radicalized form. This resurgence was driven in part by the sense of humiliation and perceived defeat that accompanied the embrace of a merely 'exhausted peace'.

'Peace as Victory': Future Visions of the Country

A third and final dimension synthesizes respondents' visions of the future and considers what these imaginaries imply for the potential for peace in Ukraine, Syria, and the Palestinian Territories. When respondents from the three groups were asked about their ideal future, they predominantly articulated 'civic' rather than 'ethnic' visions of the nation. Across all three samples, a majority expressed a desire for a democratic state grounded in an inclusive idea of citizenship, defined by shared political values rather than ethnicity, religion, or descent.

In the Ukrainian case, the ideal future envisioned by a majority of respondents is that of an independent state within the 1991 borders (85%), governed democratically (90%) and based on legal citizenship open to all who adhere to Ukraine's core national ideology. For most respondents, this 'core national ideology' entails a distinctly Western or European orientation. Moreover, 87% stated that anyone—regardless of origin, ethnicity, or religion, including Russians—can belong to the Ukrainian nation, provided they adopt this orientation. The ideal Ukrainian society, as expressed in the responses, is therefore defined primarily by political principles rather than ethnic or religious criteria. Significantly, these political orientations—centred

[3] For opinion polls among Ukrainians, see, for instance, 'The Democratic Initiatives Foundation' (DIF): https://dif.org.ua/en/category/opinion-polls (viewed 16 September 2024). For opinion polls among Palestinians, see 'The Arab Barometer': https://www.arabbarometer.org/countries/palestine/ (viewed 16 September 2024).

on liberal-democratic values and a pro-European outlook—stand in direct opposition to the political model represented by the aggressor, Russia.

In the Syrian sample, a majority of 79% describe the ideal Syria in its current national borders with the inclusion of the parts annexed by Turkey and Israel. Furthermore, the ideal state consists of a democratic government (64%) and a Syrian society based on legal citizenship for all religious and ethnic groups (48%). While for a majority of respondents, the ideal Syrian society is characterized by liberal-democratic orientations, a minority of 22% wishes for an Islamic constitution for Syria with a Sunni majority in power that rules by Islamic laws. In either vision, however, the ideal state was defined in clear opposition to the Assad government and its ideas for the country.

Finally, in the Palestinian sample, the trends are less uniform. The largest share of respondents (46%) envision the future Palestinian state within the territory of historical Palestine. Among them, a majority (60%) desire a democratic, pluralistic, and secular state. In contrast, 40% define their ideal Palestinian state as one governed by Islamic laws, with an Islamic constitution and system of government. While those who support a liberal-democratic vision based on citizenship form the majority, the Palestinian sample reveals a significant divide regarding the political character of the desired state. Importantly, regardless of whether respondents preferred a liberal-democratic or Islamic framework, all expressed a vision of Palestine's future that stood in clear opposition to their out-group, the Israelis. This oppositional framing also helps explain the strong emphasis on the Islamic character of the future state, evident in approximately 50% of responses. The ideal future for Palestine, as expressed by respondents, is a self-determined state located in historic Palestine—defined, crucially, by its distinctness from Israel.

In all three samples, a majority of respondents expressed support for a democratic state based on an inclusive conception of citizenship—one grounded in shared political principles rather than ethnicity, religion, or descent. However, as the in-depth qualitative analysis revealed, this preference was often not rooted in a strong ideological commitment to democratic values. Instead, respondents across the cases tended to adopt democratic ideals as a means of positioning their vision of the future in opposition to that of the perceived out-group. In this context, the democratic state functioned less as an intrinsic goal than as a symbolic counterpoint to the enemy.

The same pattern was evident in responses that invoked religious frameworks, particularly in the Palestinian sample. Yet here too, motivations appeared to be shaped less by religious conviction and more by the desire to define the in-group in opposition to the out-group. In this sense, both democratic and religious visions of the future functioned as political statements of differentiation rather than as expressions of intrinsic ideological commitment.

Thus, in each sample, for a majority of respondents, the ideal future is framed in exactly opposite terms to their enemies/out-groups. Whether regarding the political system, the ethnic and religious make-up, or concerning the preference for borders, the ideal country is built on political creeds that are in stark contrast to whoever is considered the enemy. The strong prevalence of liberal-democratic ideas among the Syrian sample attests to this. So does the comparatively lower embrace of liberal-democratic values in the Palestinian sample. In the Syrian case, democracy was viewed as the alternative to the authoritarian rule of Assad, whereas in the Palestinian case, liberal-democratic values resemble the Israelis and the West too closely, thus almost half of the answers desire an Islamic character for the future Palestinian state. Yet these apparent 'sectarian' designations of the state are based on political considerations vis-à-vis the out-group rather than on actual ethical or moralistic thinking. Instead of being firmly anchored in societal values, Islamic or liberal-democratic ideals are mainly deployed by respondents as political designations for the in-group to show its distinctiveness from the out-group.

In this way, the outlines of an ideal future often mirror the contours of conflict, shaped by a polarized view of the opponent as the 'other' in contrast to the 'self'. These findings suggest that for many individuals who experience war, the concept of 'peace' is synonymous with 'victory'. Yet when peace is imagined in terms of victory, the prospects for reaching a sustainable agreement diminish. In this context, the potential benefits of peace are not evaluated in terms of what it offers the in-group, but rather in terms of the perceived 'unwanted' gains it affords the out-group (Bar-Tal et al. 2009, 231). As Susan Sontag noted in her Jerusalem Prize acceptance speech: 'What most people mean by peace, of course, is victory; the victory of their own side. Calls for peace for the defeated often appear for that reason fraudulent' (Sontag 2001, quoted in Coker 2014, 85).

As Sontag's remark underscores, peace conceived as victory obstructs the willingness to compromise through negotiation. The widespread rejection of compromise—particularly regarding territorial concessions—in both Ukraine and the Palestinian Territories further illustrates this tendency (Dill et al. 2023; Pollock 2019). In the Syrian case, the notion of 'victory' materialized with unexpected speed in December 2024, when the Islamist rebel group Hay'at Tahrir al-Sham (HTS) overthrew President Bashar al-Assad and appointed their leader, Ahmed al-Sharaa, as Syria's transitional president.

However, the political vision articulated by al-Sharaa—particularly his emphasis on Islamic law as the foundation for an interim constitution—aligns more closely with the preferences of a minority of respondents (22%) who supported the establishment of an Islamic state, rather than the plurality (48%) who envisioned a liberal-democratic Syria. Although the fall of Assad's long-standing regime represents the fulfilment of a central aspiration shared by nearly all participants in this study, it is uncertain whether the political transition will ultimately reflect their broader hopes for Syria's future. Consequently, the extent to which this 'victory' can translate into a durable and inclusive peace is still an open question.

Concluding Remarks: Peace—from Idea to Practice

'Imagine Peace', printed the artist Yoko Ono on a blank page in the *New York Times* on New Year's Day of 2013. 'Imagine all the people living life in peace' many years before became her husband John Lennon's most famous songline. The call that peace must be imagined for it to become real was also echoed by IR scholars. Most prominently, John Mueller (2011) in 'War and Ideas' insists that war is only an effect of the mind, and thus people by 'an act of supreme imagination' (Coker 2014, 19) can reason themselves out of it. In the reality of war, this thought process is unfortunately not that easy.

Peace is an idea that can be imagined in many different forms. As the trends from this study's respondents showed, what most people affected by wars imagine as peace is 'victory'. In reality, such peace excludes the other side entirely and any potential compromise with it, and thus cannot create a positive peace in the long run. If anything, the high tolls of achieving that victory may eventually begin to wear off and produce the idea of an 'exhausted peace', which, once it meets reality, is hardly sustainable in the long term. In either case, the polarized dynamics of conflict stand in the way of imagining a peace that 'reconciles different values', let alone achieving it. In fact, for such peace, people and their political leaders would require rather unlikely feats of imagination: adopting a self-critical view, including the perspective of others, facing the pain, and ultimately channelling negative emotions into positive outcomes. Admittedly, this is a superhuman task, especially against the backdrop of the injustices of war.

Nevertheless, and irrespective of the outcome, imagining peace ultimately remains an endeavour that is up to the people who experience and live through war. In other words, those affected by war should be left free to imagine a peace they can live with. The trends revealed by the study of the

Ukrainian, Syrian, and Palestinian samples, as well as their individual voices, thus neither serve to instruct nor to judge people. Rather, they hope to illustrate the realities of experiencing war and the obstacles to reimagining these realities that individuals and collectives face in the wake of suffering political violence.

These insights underline that the barriers to conflict transformation remain large, yet perhaps not insurmountable. Changes in social behaviour are within the power of individuals and collectives. The framings and narratives that people, societies, and their political leaders rely upon can be altered and rewritten. The findings of this study can only assist in this endeavour: they can serve to critically review ways of thinking, collective and individual, in these and other cases of conflict around the world. Perhaps, in this way, the book can contribute to seeing war and peace in a new light. Even if only as an imagined practice of the mind, such a changed perspective bears the potential to provide everyone, readers, policymakers, and most importantly, those forced to live through political violence, with the means to wage peace instead of war.

Bibliography

Abdelal, Rawi, Herrera, Yoshiko M., Johnston, Alastair I., and McDermott, Rose (eds.) (2009). *Measuring Identity. A Guide for Social Scientists.* Cambridge: Cambridge University Press.

Acosta, Laura (2021). 'Victimhood Dissociation and Conflict Resolution: Evidence from the Colombian Peace Plebiscite', *Theory and Society*, 50(4): 679–714.

Ajami, Fouad (1978). 'The End of Pan-Arabism', *Foreign Affairs*, 57(2): 355–373, https://doi.org/10.2307/20039865.

Ajami, Fouad (1992). *The Arab Predicament: Arab Political Thought and Practice since 1967. Updated edition.* Cambridge: Cambridge University Press.

Aldoughli, Rahaf (2019). 'Interrogating the Constructions of Masculinist Protection and Militarism in the Syrian Constitution of 1973', *Journal of Middle East Women's Studies*, 15(1): 48–74.

Aldoughli, Rahaf (2022). 'What Is Syrian Nationalism? Primordialism and Romanticism in Official Ba'ath Discourse', *Nations and Nationalism*, 28(1): 1–18.

Alexander, Jeffrey C. (2012). *Trauma. A Social Theory.* Cambridge: Polity.

Alexander, Jeffrey C. (2004). 'Toward a Theory of Cultural Trauma', in: Alexander, Jeffrey C., Eyerman, Ron, Giessen, Bernhard, Smelser, Neil J., and Sztompka, Piotr. *Cultural Trauma and Collective Identity.* Berkeley: University of California Press, 1–30.

Antoniou, Georgios, Dinas, Elias, and Kosmidis, Spyros (2020). 'Collective Victimhood and Social Prejudice: A Post-Holocaust Theory of Anti-Semitism', *Political Psychology*, 41(5): 861–886.

Arel, Dominique and Driscoll, Jesse (2023). *Ukraine's Unnamed War: Before the Russian Invasion of 2022.* Cambridge: Cambridge University Press.

Aron, Raymond (1966). *Peace & War. A Theory of International Relations.* New York: Doubleday & Company.

Ayoob, Mohammed (2007). 'State Making, State Breaking, and State Failure', in: Crocker, Chester A., Hampson, Fen Osler, and Aall, Pamela (eds.). *Leashing the Dogs of War: Conflict Management in a Divided World.* Washington: United States Institute of Peace Press, 95–114.

Bachleitner, Kathrin (2021). *Collective Memory in International Relations.* Oxford: Oxford University Press.

Bachleitner, Kathrin (2022). 'Legacies of War. Syrian Narratives of Conflict and Visions of Peace', *Cooperation and Conflict*, 57(1): 43–64.

Bachleitner, Kathrin (2023). 'Ontological Security as Temporal Security? The Role of "Significant Historical Others" in World Politics', *International Relations*, 37(1): 25–47.

Bachleitner, Kathrin (2024). 'Trauma in World Politics: Memory Dynamics Between Different Victim Groups', *Journal of Peace Research*, 62(3): 629–642.

Bachleitner, Kathrin and Betts, Alexander (2024). 'The EU's Normative Dissensus on Migration. How National Identities Shaped Responses to the European Refugee Crisis', *Journal of European Integration*, 47(3): 469–484.

Bachleitner, Reinhard, Weichbold, Martin, and Wolfgang Aschauer (2010). *Die Befragung im Kontext von Raum, Zeit und Befindlichkeit: Beiträge zu einer prozessorientierten Theorie der Umfrageforschung.* Wiesbaden: VS Verlag.

Baker, William D. and Oneal, John R. (2001). 'Patriotism or Opinion Leadership?: The Nature and Origins of the "Rally 'Round the Flag" Effect', *The Journal of Conflict Resolution*, 45(5): 661–687.

Balcells, Laia (2012). 'The Consequences of Victimization on Political Identities: Evidence from Spain', *Politics and Society*, 40(3): 311–347.

Bar-Tal, Daniel (2007). 'Sociopsychological Foundations of Intractable Conflicts', *American Behavioral Scientist*, 50(11): 1430–1453.

Bar-Tal, Daniel (2013). *Intractable Conflicts: Socio-Psychological Foundations and Dynamics*. Cambridge: Cambridge University Press.

Bar-Tal, Daniel, Halperin, Eran, Porat, Roni, and Nets-Zehngut, Roni (2013). 'Why Society Members Tend to Support the Continuation of Intractable Conflicts and Resist Peaceful Resolution', in: Golec de Zavala, Agnieszka and Cichocka, Aleksandra (eds.). *Social Psychology of Social Problems: The Intergroup Context*. London: Palgrave Macmillan/Springer Nature, 254–279.

Bar-Tal, Daniel, Chernyak-Hai, Lily, Schori, Noa, and Gundar, Ayelet (2009). 'A Sense of Self-perceived Collective Victimhood in Intractable Conflicts', *International Review of the Red Cross*, 91(874): 229–258.

Bar-Tal, Daniel, Halperin, Eran, and Oren, Neta (2010). 'Socio-psychological Barriers to Peace Making: The Case of the Israeli Jewish Society', *Social Issues and Policy Review*, 4(1): 63–109.

Bar-Tal, Daniel and Hammack, Phillip L. (2012). 'Conflict, Deligitimization, and Violence', in: Tropp, Linda R. (ed.). *The Oxford Handbook of Intergroup Conflict*. New York: Oxford University Press, 29–52.

Bar-Tal, Daniel, Oren, Neta, and Nets-Zehngut, Rafi (2014). 'Sociopsychological Analysis of Conflict-Supporting Narratives: A General Framework', *Journal of Peace Research*, 51(5): 662–675.

Barnett, Michael N. (1995). 'Sovereignty, Nationalism, and Regional Order in the Arab States System', *International Organization*, 49(3): 479–510.

Bates, Robert H. (2008). 'State Failure', *Annual Review of Political Science*, 11: 1–12.

Bauer, M., Blattman, C., Chytilová, J., Henrich, J., Miguel, E., and Mitts, T. (2016). 'Can war foster cooperation?', *Journal of Economic Perspectives*, 30(3): 249–74.

Bechhofer, F., McCrone, D. (2009). 'National Identity, Nationalism and Constitutional Change', in: Bechhofer, F. and McCrone, D. (eds.). *National Identity, Nationalism and Constitutional Change*. London:Palgrave Macmillan, 1–16.

Bechhofer, Frank and McCrone, David (eds.) (2009). *National Identity, Nationalism and Constitutional Change*. London:Palgrave MacMillan.

Bell, Duncan (ed.) (2010). *Memory, Trauma and World Politics. Reflections of the Relationship between Past and Present*. Paperback edition. London: Palgrave Macmillan.

Bellows, J. and Miguel, E. (2009). 'War and Local Collective Action in Sierra Leone', *Journal of Public Economics*, 93(11): 1144–1157.

Benhabib, Seyla, Shapiro, Ian, and Petranovic, Danilo (eds.) (2007). *Identities, Affiliations and Allegiances*. Cambridge: Cambridge University Press.

Berenskoetter, Felix (2023). 'Memory, Identity, and Its Politics', in: Mälksoo, Maria (ed.). *Handbook on the Politics of Memory*. Cheltenham: Edward Elgar Publishing, 18–30.

Berenskoetter, Felix (2014). 'Parameters of a National Biography', *European Journal of International Relations*, 20(1): 262–288.

Betts, Alexander and Loescher, Gil (eds.) (2010). *Refugees in International Relations*. Oxford: Oxford University Press.

Blainey, Geoffrey (1988). *Causes of War*. London: Palgrave MacMillan.

Bose, Sumantra (2007). *Contested Lands. Israel-Palestine, Kashmir, Bosnia, Cyprus, and Sri Lanka*. London: Harvard University Press.

Branscombe, Nyla R. (2004). 'A Social Psychological Process Perspective on Collective Guilt', in: Branscombe, Nyla R. and Doosje, Bertjan. *Collective Guilt. International Perspectives.* Cambridge: Cambridge University Press, 320–334.

Brighton, Shane (2011). 'Three Propositions on the Phenomenology of War', *International Political Sociology*, 5(1): 101–105.

Browers, Michaelle (2009). *Political Ideology in the Arab World: Accommodation and Transformation.* Cambridge: Cambridge University Press.

Brubaker, Rogers (1992). *Citizenship and Nationhood in France and Germany.* Cambridge, MA: Harvard University Press.

Bunton, Martin (2013). *The Palestinian–Israeli Conflict. A Very Short Introduction.* Oxford: Oxford University Press.

Butler, Judith (2004). *Precarious Life. The Power of Mourning and Violence.* London: Verso.

Calhoun, Craig (1992). 'Nationalism and Ethnicity', *Annual Review of Sociology*, 19(1): 211–239.

Calhoun, Craig (2007). 'Social Solidarity as a Problem for Cosmopolitan Democracy', in: Benhabib, Seyla, Shapiro, Ian, and Petranovic, Danilo (eds.). *Identities, Affiliations and Allegiances.* Cambridge: Cambridge University Press, 285–302.

Canetti, Daphna, Elad-Strenger, Julia, Lavi, Iris, Guy, Dana, and Daniel Bar-Tal (2015). 'Exposure to Violence, Ethos of Conflict, and Support for Compromise: Surveys in Israel, East Jerusalem, West Bank, and Gaza', *Journal of Conflict Resolution*, 61(1): https://doi.org/10.1177/0022002715569771.

Caselli, Francesco and Coleman, Wilbur John (2006). 'On the Theory of Ethnic Conflict', *National Bureau of Economic Research NBER Working Paper* No. w12125. Available at SSRN: https://ssrn.com/abstract=893780.

Cashman, Greg (2013). *What Causes War? An Introduction to Theories of International Conflict.* Lanham, MD: Rowman & Littlefield Publishers.

Cederman, Lars-Erik and Gleditsch, Kristian S. (2013). *Inequality, Grievances, and Civil War.* Cambridge: Cambridge University Press.

Cederman, Lars-Erik, Camber,Warren T., and Sornette, Didier (2011). 'Testing Clausewitz: Nationalism, Mass Mobilization, and the Severity of War', *International Organization*, 65(4): 605–638.

Chaitin, Julia and Avlagon, Eldad (2025). *Striving for Peace through Personal Narratives of Genocide and War.* Cambridge: Cambridge University Press.

Chevée, Adelie (2022). 'From Suriyya al-Asad to Souriatna: Civic Nationalism in the Syrian Revolutionary Press', *Nations and Nationalism*, 28(1): 154–176.

Cockell, John G. (2001). 'Human Security and Preventive Action Strategies', in: Newman, Edward and Oliver P. Richmond (eds.). *The United Nations and Human Security.* London: Palgrave MacMillan, 15–30.

Coker, Christopher (2014). *Can War Be Eliminated?* Cambridge: Polity Press.

Collier, Paul and Hoeffler, Anke (1998). 'On Economic Causes of Civil War', *Oxford Economic Papers*, 50(4): 563–573.

Confino, Alon (1997). 'Collective Memory and Cultural History: Problems of Method', *The American Historical Review*, 102(5): 1386–1403.

Coser, Lewis A. (1956). *The Functions of Social Conflict.* New York: Free Press.

Cruz, Consuelo (2000). 'Identity and Persuasion: How Nations Remember Their Pasts and Make Their Futures', *World Politics*, 52(3): 275–312.

Dahrendorf, Ralph (1958). 'Toward a Theory of Social Conflict', *Journal of Conflict Resolution*, 2(2): 170–183

Dawisha, Adeed (2000). 'Arab Nationalism and Islamism: Competitive Past, Uncertain Future', *International Studies Review*, 2(3): 79–90.

Dill, Janina, Howlett, Marnie, and Müller-Crepon, Carl (2023). 'At Any Cost: How Ukrainians Think About Self-Defense Against Russia', *American Journal of Political Science*, 68(4): 1460–1478: https://doi.org/10.1111/ajps.12832.

Ditlmann, Ruth, Purdie-Vaughns, Valerie, Dovidio, John F., and Naft, Michael J. (2017). 'The Implicit Power Motive in Intergroup Dialogues About the History of Slavery', *Journal of Personality and Social Psychology*, 112(1): 116–135.

Dyrstad, Karin (2012). 'After Ethnic Civil War: Ethno-Nationalism in the Western Balkans', *Journal of Peace Research*, 49(6): 817–831.

Dyrstad, Karin, Binningsbø, Helga M., and Bakke, Kristin M. (2022). 'Wartime Experiences and Popular Support for Peace Agreements: Comparative Evidence from Three Cases', *Journal of Conflict Resolution*, 66(9): 1562–1588.

Dyrstad, Karin, Buhaug, Halvard, Ringdal, Kristen, Simkus, Albtert, and Listhaug, Ola (2011). 'Microfoundations of Civil Conflict Reconciliation: Ethnicity and Context', *International Interactions*, 37(4): 363–387.

Edkins, Jenny (2002). 'Forget Trauma? Responses to September 11', *International Relations*, 16 (2): 243–256.

Edkins, Jenny (2003). *Trauma and the Memory of Politics.* Cambridge: Cambridge University Press.

Eisenstadt, Shmuel N. and Giesen, Bernhard (1995). 'The Construction of Collective Identity', *European Journal of Sociology*, 36 (1): 72–102.

Elshtain, Jean Bethke (1987). *Women and War.* New York: Basic Books.

Erikson, Kai R. (1976). *Everything in Its Path: Destruction of Community in the Buffalo Creek Flood.* New York: Simon and Schuster.

Farouk-Alli, Aslam (2014). 'Sectarianism in Alawi Syria: Exploring the Paradoxes of Politics and Religion', *Journal of Muslim Minority Affairs*, 34(3): 207–226.

Fearon, James D. (1995). 'Rationalist Explanations for War', *International Organization*, 49(3): 379–414.

Fierke, Karin (2004). 'Whereof We Can Speak, Thereof We Must Not Be Silent: Trauma Political Solipsism and War', *Review of International Studies*, 30(4): 471–491.

Fierke, Karin M. (2012). *Political Self-Sacrifice: Agency, Body and Emotion in International Relations.* Cambridge: Cambridge University Press.

Firchow, Pamina (2018). *Reclaiming Everyday Peace: Local Voices in Measurement and Evaluation After War.* Cambridge: Cambridge University Press.

Freedman, Lawrence (2012). 'Defining War', in: Boyer, Yves and Lindley-French, Julian (eds.). *The Oxford Handbook of War.* Oxford: Oxford University Press, 16–29.

Fujii, Lee Ann (2012). 'Research Ethics 101: Dilemmas and Responsibilities', *Political Science and Politics*, 45(4): 717–723.

Fukuyama, Francis (2018). *Identity: The Demand for Dignity and the Politics of Resentment.* London: Profile Books.

Galtung, Johan (1969). 'Violence, Peace and Peace Research', *Journal of Peace Research*, 6(3): 167–191.

Gibler, Douglas M., Hutchison, Marc L., and Miller, Steven V. (2012). 'Individual Identity Attachments and International Conflict: The Importance of Territorial Threat: *Comparative Political Studies*', 45(12): 1655–1683

Gorenberg, Gershom (2006). *The Accidental Empire. Israel and the Birth of the Settlements, 1967–77.* New York: Times Books.

Gray, Colin S. (2012). *War, Peace and International Relations: An Introduction to Strategic History.* London: Routledge.

Greig, Michael (2008). 'When Do They Say Yes? An Analysis of the Willingness to Offer and Accept Mediation in Civil Wars', *International Studies Quarterly*, 52(4): 759–781.

Hall, John A. and Malešević, Sinisa (2013). 'Introduction. Wars and Nationalisms', in: Hall, John A. and Malešević, Sinisa (eds.). *Nationalism and War*. Cambridge: Cambridge University Press, 1–28.
Haran, Olexiy, Yakovlyev, Maksym, and Zolkina, Maria (2019). 'Identity, War, and Peace: Public Attitudes in the Ukraine-Controlled Donbas', *Eurasian Geography and Economics*, 60(6): 684–708.
Hardin, Russell (1995). *One for All: The Logic of Group Conflict*. Princeton: Princeton University Press.
Hartman, Alexandra C. and Morse, Benjamin S. (2018). 'Violence, Empathy and Altruism: Evidence from the Ivorian Refugee Crisis in Liberia', *British Journal of Political Science*, 50(2): 731–755.
Hassner, Ron E. (2009). *War on Sacred Grounds*. Ithaca: Cornell University Press.
Herbst J. (1990). 'War and the State in Africa', *International Security*, 14(4): 117–139.
Hesse, Hermann (1946). *Krieg und Frieden. Betrachtungen zu Krieg und Politik seit dem Jahr 1914*. English Translation: *If the War Goes On* (2018). Edinburgh: Canongate Books.
Hiers, W. and Wimmer, Andreas (2013). 'Is Nationalism the Cause or Consequence of Imperial Breakdown', in: Hall, S. and Malešević, S. (eds.). *Nationalism and War*. Cambridge: Cambridge University Press, 212–254.
Hinnebusch, Raymond (1989). *Peasant and Bureaucracy in Ba'thist Syria: The Political Economy of Rural Development*. Boulder, CO: Westview Press.
Hinnebusch, Raymond (2020a). 'Identity and State Formation in Multi-sectarian Societies: Between Nationalism and Sectarianism in Syria', *Nations and Nationalism*, 26(1): 138–154.
Hinnebusch, Raymond (2020b). 'The Battle over Syria's Reconstruction', *Global Policy*, 11(1): 113–123.
Hokayem, Emile (2017). *Syria's Uprising and the Fracturing of the Levant*. Abingdon: Routledge.
Hosli, Madeleine, and Hoekstra, Anke (2013). 'What Fosters Enduring Peace? An Analysis of Factors Influencing Civil War Resolution', *Peace Economics, Peace Science and Public Policy*, 19(2): 123–155
Howlett, Marnie (2023a). 'The Russia-Ukraine War as a Battle for a Bordered Land, Not Borderland', *Political Geography*, 101(102814): https://doi.org/10.1016/j.polgeo.2022.102814
Howlett, Marnie (2023b). 'Nation-Building from (Below) the Grassroots: Everyday Nationalism in Ukraine's Bomb Shelters', *Nations and Nationalism*, 29(1): 63–69.
Huddy, Leonie, Del Ponte, Alessandro, and Davies, Caitlin (2021). 'Nationalism, Patriotism, and Support for the European Union', *Political Psychology*, 42(6): 995–1017.
Hunt, Nigel (2010). *Memory, War and Trauma*. Cambridge: Cambridge University Press.
Hutchison, Emma (2016). *Affective Communities in World Politics*. Cambridge: Cambridge University Press.
Hutchison, John (2017). *Nationalism and War*. Oxford: Oxford University Press.
Ismail, Salwa (2011). 'The Syrian Uprising: Imagining and Performing the Nation', *Studies in Ethnicity and Nationalism*, 11(3): 538–549.
Jawad, Saleh A. (2006). 'The Arab and Palestinian Narratives of the 1948 War', in: Rotberg, Robert I. (ed.). *Israeli and Palestinian Narratives of Conflict. History's Double Helix*. Bloomington: Indiana University Press, 72–114.
Kaderan, Burak (2017). 'Nationalism and War for Territory: From 'Divisible' Territories to Inviolable Homelands', *Cambridge Review of International Affairs*, 30(4): 368–393.
Käihkö, Ilmari (2022). 'The Utility of Ethnography for Understanding (the Russo-Ukrainian) War', *Journal of Ethnographic Theory*, 12(3): 677–685.
Kalyvas, Stathis (2006). *The Logic of Violence in Civil War*. Cambridge: Cambridge University Press.

Kalyvas, Stathis and Sambanis, Nicholas (2005). 'Bosnia's Civil War: Origins and Violence Dynamics', in: Collier, Paul and Sambanis, Nicholas (eds.). *Understanding Civil War: Evidence and Analysis.* Washington, DC: The World Bank, Volume 2, 191–229.

Kastoryano, Riva (2007). 'Transnational Nationalism: Redefining Nations and Territory', in: Seyla Benhabib, Ian Shapiro and Danilo Petranovic (eds.). *Identities, Affiliations and Allegiances.* Cambridge: Cambridge University Press, 159–180.

Kaufmann, Chaim D. (1996a). 'Intervention in Ethnic and Ideological Civil Wars: Why One Can Be Done and the Other Can't', *Security Studies*, 6(1): 62–100.

Kaufmann, Chaim D. (1996b). 'Possible and Impossible Solutions to Ethnic Civil Wars', *International Security*, 20(4): 136–175.

Kaufmann, Stuard J. (2001). *Modern Hatreds: The Symbolic Politics of Ethnic War.* London: Cornell University Press.

Kelman, Herbert C. (1987). 'The Political Psychology of the Israeli-Palestinian Conflict: How Can We Overcome the Barriers to a Negotiated Solution?', *Political Psychology* 8(3): 347–363.

Kinnvall, Caterina (2013). 'Trauma and the Politics of Fear: Europe at the Crossroads', in: Demertzis, Nicholas (ed.), *Emotions in Politics: The Affect Dimension in Political Tension.* Basingstoke: Palgrave Macmillan, 143–166.

Kulyk, Volodymyr (2016). 'National Identity in Ukraine: Impact of Euromaidan and the War', *Europe-Asia Studies*, 68(4): 588–608.

Kurzman, Charles (2004). 'Can Understanding Undermine Explanation? The Confused Experience of Revolution', *Philosophy of the Social Sciences*, 34(3): 328–351.

Laitin, James D. (2007). *Nations, States and Violence.* Oxford: Oxford University Press.

Lake, David A. and Rothchild, Donald (1996). 'Containing Fear: The Origins and Management of Ethnic Conflict', *International Security*, 21(2): 41–75.

Lebow, Richard N. (2016). *National Identities and International Relations.* Cambridge: Cambridge University Press.

Lerner, Adam (2020). 'The Uses and Abuses of Victimhood Nationalism in International Politics', *European Journal of International Relations*, 26(1): 62–87.

Levy, Jack S. and Thompson, William R. (2010). *Causes of War.* Chichester, UK: Wiley-Blackwell.

Lupu, Noam and Peisakhin, Leonid (2017). 'The Legacy of Political Violence Across Generations', *American Journal of Political Science*, 61(4): 836–851.

Mac Ginty, Roger (2021). *Everyday Peace: How So-Called Ordinary People Can Disrupt Violent Conflict.* New York: Oxford University Press.

MacMillan, Margaret (2021). *War: How Conflict Shaped Us.* London: Random House Trade Paperbacks.

Malešević, Sinisa (2011a). 'Nationalism, War and Social Cohesion', *Ethnic and Racial Studies* 34(1): 141–61.

Malešević, Sinisa (2011b). 'The Chimera of National Identity', *Nations and Nationalism*, 17(2): 272–290.

Manekin, Devorah, Grossman, Guy, and Mitts, Tamar (2018). 'Contested Ground: Disentabling Material and Symbolic Attachment to Disputed Territory', *Political Science Research and Methods*, 7(4): 679–697.

Mann, Michael (2013). 'The Role of Nationalim in the Two World Wars', in: Hall, John A. and Malešević, Sinisa (eds.). *Nationalism and War.* Cambridge: Cambridge University Press, 172–196.

Maoz, Ifat, and McCauley, Clark (2005). 'Psychological Correlates of Support for Compromise: A Polling Study of Jewish-Israeli Attitudes toward Solutions to the Israeli-Palestinian Conflict', *Political Psychology*, 26(5): 791–807.

Martín, Félix E. (2005). 'Critical Analysis of the Concept of Peace in International Relations', *Peace Research*, 37(2): 45–59.

Matthews, Bob and Ross, Liz (2010). *Research Methods. A Practical Guide for the Social Sciences*. London: Pearson Education Limited.

McCann, Lisa I., and Pearlman, Laurie Anne (1990). 'Vicarious Traumatization: A Framework for Understanding the Psychological Effects of Working with Victims', *Journal of Traumatic Stress*, 3(1): 131–149.

McNabb, David E. (2021). *Research Methods For Political Science. Quantitative, Qualitative and Mixed Method Approaches*. Third edition. New York: Taylor & Francis.

Mearsheimer, John (1995). 'The False Promise of International Security', *International Security*, 19(3): 5–49.

Meinecke, Friedrich. (1970)/(1907). *Cosmopolitanism and the National State*. Princeton: Princeton University Press.

Mitzen, Jennifer (2006). 'Ontological Security in World Politics: State Identity and the Security Dilemma', *European Journal of International Relations*, 12(3): 341–370.

Monroe, Kristen R. (2011). *Ethics in an Age of Terror and Genocide: Identity and Moral Choice*. Princeton: Princeton University Press.

Morgenthau, Hans (1970). *Truth and Power: Essays of a Decade, 1960–70*. New York: Praeger.

Mueller, John (2011). *War and Ideas*. London: Routledge.

Neal, Arthur G. (1998). *National Trauma and Collective Memory*. New York: Routledge.

Nemeroff, Charles B. and Marmar, Charles R. (eds.) (2018). *Post-Traumatic Stress Disorder*. New York: Oxford University Press.

Newman, Edward and Richmond, Oliver P. (2001). 'Introduction: Beyond Peacekeeping', in: Newman, Edward and Richmond, Oliver P. (eds.) *The United Nations and Human Security*. London: Palgrave MacMillan, 3–14.

Newman, Edward and Richmond, Oliver P. (eds.) (2001). *The United Nations and Human Security*. London: Palgrave MacMillan.

Noor, Masi, Vollhardt, Johanna R., Mari, Silvia, and Nadler, Arie (2017). 'The Social Psychology of Collective Victimhood', *European Journal of Social Psychology*, 47(2): 121–134.

Nussio, Enzo, Rettberg, Angelika, and Ugarriza, Juan E. (2015). 'Victims, Nonvictims and Their Opinions on Transitional Justice: Findings from the Colombian Case', *International Journal of Transitional Justice*, 9(2): 336–354.

Olick, Jeffrey K. (1998). 'Introduction: Memory and the Nation: Continuities, Conflicts, and Transformations', *Social Science History*, 22(4), Special Issue: 'Memory and the Nation': 377–387.

Olick, Jeffrey K. (2005). *In the House of the Hangman: The Agonies of German Defeat, 1943–1949*. Chicago/London: University of Chicago Press.

Olick, Jeffrey K. and Robbins, Joyce (1998). 'Social Memory Studies: From "Collective Memory" to the Historical Sociology of Mnemonic Practices', *Annual Review of Sociology*, 241(1): 105–140.

Olick, Jeffrey K., Vinitzky-Seroussi, Vered, and Levy, Daniel (eds.) (2011). *The Collective Memory Reader*. Oxford: Oxford University Press.

Olmert, Josef (2011). 'Israel–Syria: The Elusive Peace', *Domes*, 20(2): 202–211.

Onuch, Olga and Sasse, Gwendolyn (2016). 'The Maidan in Movement: Diversity and the Cycles of Protest', *Europe-Asia Studies*, 68(4): 556–587.

Onuch, Olga, and Hale, Henry E. (2022). *The Zelensky Effect*. London: Hurst.

Oren, Michael B. (2002). *Six Days of War. June 1967 and the Making of the Modern Middle East*. Oxford: Oxford University Press.

Parekh, Bhikhu (1995). 'The Concept of National Identity', *Journal of Ethnic and Migration Studies*, 21(2): 255–268.

Pearlman, Wendy (2016). 'Narratives of Fear in Syria', *Perspectives on Politics*, 14(1): 21–37.
Pearlman, Wendy (2017). 'Memory as a Field Site', *International Journal of Middle East Studies*, 49(3): 501–505.
Pearlman, Wendy (2023). 'Emotional Sensibility: Exploring the Methodological and Ethical Implications of Research Participants' Emotions', *American Political Science Review*, 117(4): 1241–1254.
Perthes, Volker. (2004). *Syria Under Bashar al-Asad: Modernisation and the Limits of Change*. Oxford: Oxford University Press.
Petersen, Roger D. (2002). *Understanding Ethnic Violence: Fear, Hatred, and Resentment in Twentieth-Century Eastern Europe*. New York: Cambridge University Press.
Phillips, Christopher (2013). *Everyday Arab Identity: The Daily Reproduction of the Arab World*. London: Routledge.
Phillips, Christopher (2015). 'Sectarianism and Conflict in Syria', *Third World Quarterly*, 36(2): 357–376.
Phillips, Christopher (2017). *The Battle for Syria*. New Haven: Yale University Press.
Pires, Edmundo B., Carvalho, Claudio A. S., and Joana Ricarte (eds.) (2024). *Memory, Trauma and Narratives of the Self. An Interdisciplinary Approach*. Cheltenham, UK: Edgar Elgar.
Pisano, Jessica (2022). *Staging Democracy: Political Performance in Ukraine, Russia, and Beyond*. Ithaka: Cornell University Press.
Pollock, David (2019). 'Half of Palestinians Still Want All of Palestine—But Most Would Compromise for Less', Policy Analysis of the Washington Institute for Near East Policy, 13 August. https://www.washingtoninstitute.org/policy-analysis/half-palestinians-still-want-all-palestine-most-would-compromise-less.
Quillian, Lincoln (1995). 'Prejudice as a Response to Perceived Group Threat: Population Composition and Anti-Immigrant and Racial Prejudice in Europe', *American Sociological Review*, 60(4): 586–611.
Reeskens, Tim and Hooghe, Marc (2010). 'Beyond the Civic–Ethnic Dichotomy: Investigating the Structure of Citizenship Concepts Across Thirty-Three Countries', *Nations and Nationalism*, 16(4): 579–597.
Renan, Ernest (1882). *What Is a Nation?*, text of a speech delivered at the Sorbonne on 11 March 1882.
Ringmar, Erik (1996). 'On the Ontological Status of the State', *European Journal of International Relations*, 2(4): 439–466.
Risse-Kappen, Thomas (2010). *A Community of Europeans? Transnational Identities and Public Spheres*. Ithaca: Cornell University Press.
Rosoux, Valerie (2020). 'Memory, Cultural Heritage and Legacies of Wars', in: Osler Hampson, Fen, Özerdem, Alpaslan, and Kent, Jonathan (eds.). *Routledge Handbook of Peace, Security, and Development*. London: Routledge, 226–238.
Rotberg, Robert I. (2004). *The Failure and Collapse of Nation-States: Breakdown, Prevention, and Repair*, in: Rotberg, Robert I. (ed.). *When States Fail: Causes and Consequences*. Princeton: Princeton University Press, 1–49.
Rotberg, Robert I. (ed.) (2006). *Israeli and Palestinian Narratives of Conflict. History's Double Helix*. Bloomington: Indiana University Press.
Rotberg, Robert I. (2006). 'Preface', in: Rotberg, Robert I. (ed.). *Israeli and Palestinian Narratives of Conflict. History's Double Helix*. Bloomington, IN: Indiana University Press, vii–viii.
Rumelili, Bahar and Todd, Jennifer (2018). 'Paradoxes of Identity Change: Integrating Macro, Meso, and Micro Research on Identity in Conflict Processes', *Politics*, 38(1): 3–18.
Saito, Hiro (2006). 'Reiterated Commemoration: Hiroshima as National Trauma', *Sociological Theory*, 24(4): 353–376.

Sambanis, Nicholas, and Lee, Amber (2023). 'Does External Threat Bring the Nation Together? (Working Paper)': Available at: http://dx.doi.org/10.2139/ssrn.4440616.
Scarry, Elaine (1985). *The Body in Pain: The Making and Unmaking of the World.* Oxford: Oxford University Press.
Schaebler, Birgit (2013). 'Constructing an Identity between Arabism and Islam: The Druzes in Syria', *The Muslim World*, 103(1): 62–79.
Schatz, Edward (ed.) (2009). *Political Ethnography: What Immersion Contributes to the Study of Power.* Chicago, IL: University of Chicago Press.
Schick, Kate (2011). 'Acting Out and Working Through: Trauma and (In)Security', *Review of International Studies*, 37(4): 1837–1855.
Shlaim, Avi (1995). 'The Debate About 1948', *International Journal of Middle East Studies*, 27(3): 287–304.
Shlaim, Avi (2014). *The Iron Wall: Israel and the Arab World.* New York: Norton.
Shulman, Stephen (2002). 'Challenging the Civic/Ethnic and West/East Dichotomies in the Study of Nationalism', *Comparative Political Studies*, 35(5): 554–585.
Simmel, Georg (1955/1908). *Conflict and the Web of Group Affiliations.* Glencoe: Free Press.
Smelser, Neil J. (2004). 'Psychological Trauma and Cultural Trauma', in: Alexander, Jeffrey C., Eyerman, Ron, Giesen, Bernard, Smelser, Neil J., and Sztompka, Piotr. *Cultural Trauma and Collective Identity.* Berkeley: University of California Press, 31–59.
Smith, Anthony D. (2000). 'The "Sacred" Dimension of Nationalism', *Millennium: Journal of International Studies*, 29(3): 791–814.
Somers, Margaret R. (1994). 'The Narrative Constitution of Identity: A Relational and Network Approach', *Theory and Society*, 23(5): 605–649.
Sosnowski, Marika (2023). *Redefining Ceasefires: Wartime Order and Statebuilding in Syria.* Cambridge: Cambridge University Press.
Steele, Brent J. (2008). *Ontological Security in International Relations: Self-identity and the IR State.* London: Routledge.
Sue, Valerie M. and Ritter, Lois A. (2016). *Conducting Online Surveys: 2nd Edition.* Los Angeles: SAGE.
Sumner, William Graham (1906). *Folkways.* Boston: Ginn.
Sylvester, Christine (2012). 'War Experiences/War Practices/War Theory', *Millennium: Journal of International Studies*, 40(3): 480–503.
Sylvester, Christine (2013a). *War as Experience: Contributions from International Relations and Feminist Analysis.* London: Routledge.
Sylvester, Christine (2013b). 'Experiencing War: A Challenge for IR', *Cambridge Review of International Affairs*, 26(4): 669–674.
Tedeschi, Richard G., and Calhoun, Lawrence G. (1996). 'The Posttraumatic Growth Inventory: Measuring the Positive Legacy of Trauma', *Journal of Traumatic Stress*, 9(3): 455–471.
Tessler, Mark. 2009. *A History of the Israeli–Palestinian Conflict. 2nd ed.* Bloomington: Indiana University Press.
Tilly, Charles (1990). *Coercion, Capital and European States: AD 990–1990.* Oxford: Blackwell.
Todd, Jennifer (2018). 'The Politics of Identity Change and Conflict: An Agenda for Research', *Politics*, 38(1): 84–93.
Todd, Jennifer (2018a). *Identity Change After Conflict. Ethnicity, Boundaries and Belonging in the Two Irelands.* London: Palgrave Macmillan.
Toft, Monica Duffy (2003). *The Geography of Ethnic Violence: Identity, Interests, and the Indivisibility of Territory.* Princeton, NJ: Princeton University Press.
Tönnies, Ferdinand (1955). *Community and Association (Gemeinschaft und Gesellschaft).* London: Routledge and Kegan Paul.

Valadez, Jorge M. (2007). 'The Continuing Significance of Ethnocultural Identity', in: Benhabib, Seyla, Shapiro, Ian, and Petranovic, Danilo (eds.). *Identities, Affiliations and Allegiances*. Cambridge: Cambridge University Press, 303–324.

Van Hear, Nicholas and Cohen, Robin (2017). 'Diasporas and Conflict: Distance, Contiguity and Spheres of Engagement', *Oxford Development Studies*, 45(2): 171–184.

Villamil, Francisco (2021). 'Mobilizing Memories: The Social Conditions of the Long-Term Impact of Victimization', *Journal of Peace Research*, 58(3): 399–416.

Vinitzky-Seroussi, Vered (2002). 'Commemorating a Difficult Past: Yitzhak Rabin's Memorials', *American Sociological Review*, 67(1): 30–51.

Viroli, Maurizio (1997). *For the Love of Country. An Essay on Patriotism and Nationalism*. Oxford: Oxford University Press.

Vollhardt, Johanna R. and Bilewicz, Michal (2013). 'After the Genocide: Psychological Perspectives on Victim, Bystander and Perpetrator Group', *Journal of Social Issues*, 69(1): 1–15.

Von Clausewitz, Carl (1976). *On War*. Princeton: Princeton University Press.

Waltz, Kenneth (1959). *Man, the State, and War*. New York: Columbia University Press.

Waltz, Kenneth (1979). *Theory of International Politics*. Long Grove, IL: Waveland Press.

Wedeen, Lisa (1999). *Ambiguities of Domination: Politics, Rhetoric, and Symbols in Contemporary Syria*. Chicago: University of Chicago Press.

Wertsch, James V. (2008). 'The Narrative Organization of Collective Memory', *Ethos*, 36(1): 120–135.

White, Hayden (1987). *The Content of the Form: Narrative Discourse and Historical Representation*. Baltimore: Johns Hopkins University Press.

Wilson, Andrew (2016). 'The Donbas in 2014: Explaining Civil Conflict Perhaps, but Not Civil War', *Europe-Asia Studies*, 68(4): 631–652.

Wimmer, Andreas (2014). 'War', *The Annual Review of Sociology*, 40: 173–197.

Wimmer, Andreas (2019). 'Why Nationalism Works', *Foreign Affairs*, 98(2): 27–34.

Wimmer, Andreas and Min, Brian (2009). 'The Location and Purpose of Wars Around the World: A New Global Dataset, 1816–2001', *International Interactions*, 35(4): 390–417.

Wood, Elisabeth, J. (2008). 'The Social Processes of Civil War: The Wartime Transformation of Social Networks', *Annual Review of Political Science*, 11(1): 539–561.

Wood, Elisabeth J. (2012). *Insurgent Collective Action and Civil War in El Salvador*. Cambridge: Cambridge University Press.

Zeitzoff, Thomas (2018). 'Anger, Legacies of Violence, and Group Conflict: An Experiment in Post-Riot Acre, Israel', *Conflict Management and Peace Science*, 35(4): 402–423.

Zelensky, Volodymyr (2022). *A Message from Ukraine*. London: Hutchinson Heinemann.

Zerubavel, Eviatar (2003). *Time Maps: Collective Memory and the Social Shape of the Past*. Chicago: University of Chicago Press.

Ziser, Eyal (2001). *Asad's Legacy: Syria in Transition*. London: Hurst & Company.

Zitek, Emily M., Jordan, Alexander H., Monin, Benoît, and Leach, Frederick R. (2010). 'Victim Entitlement to Behave Selfishly', *American Psychological Association*, 98(2): 245–255.

Index

For the benefit of digital users, indexed terms that span two pages (e.g., 52–53) may, on occasion, appear on only one of those pages.

Figures are indicated by an italic *f* following the paragraph number.